Integrating Occupational Safety and Health using Digital Technologies

This book is designed to help practitioners understand the growing potential for the use of Building Information Modelling (BIM) to improve Occupational Safety and Health (OSH) outcomes in the Architecture, Engineering, Construction, and Operations sector. Through a series of cutting-edge practical cases, an international team of researchers and practitioners examine the benefits, barriers, and challenges associated with the introduction of new digital information technologies in OSH settings. Covering training, risk identification, site planning, site monitoring, emergency planning, and accident investigation, the book shows how the use of BIM paves the way for better streamlining of prevention planning, by making it more visible and understandable. The BIM applications presented allow for OSH to be closely linked to production, thus creating the conditions for an effective integration of OSH within the construction processes associated with building, operations, and management.

The book introduces BIM as an enabler of improved OSH performance in construction projects. Precise BIM applications are presented, which are directly related to OSH performance. Backed by research, the BIM applications are presented in a practical manner, making them easily applicable by the construction project professional, as well as informative for the academic researcher.

Manuel Tender (PhD)
Adjunct Professor at ISEP-Instituto Superior de Engenharia do Porto and at ISLA Polytechnic Institute of Management and Technology. Researcher at CONSTRUCT-Gequaltec, Faculty of Engineering, University of Porto. CDM Coordinator since 2001. Coordinator of the "Safety and Health" Working Group of Technical Commission 197 (BIM). Lead Researcher at the Digital4OSH Project.

Paul Fuller (ENGD)
Visiting Academic at the School of Architecture, Building and Construction Management of Loughborough University. Session Lecturer in Construction Project Management. Forty years of international experience in both large- and small-scale project environments.

Peter Demian (PhD)
MCIOB. Reader in Loughborough University. Professor of Digital Construction Engineering and Leader of Construction Management Group.

Matej Mihic (PhD)
Assistant Professor at the University of Zagreb – Faculty of Civil Engineering in the Department of Construction Management, teaching courses related to construction safety, construction management, and construction technologies. President of the Croatian Association of Health and Safety Coordinators in Construction.

João Pedro Couto (PhD)
Assistant Professor in the Civil Engineering Department at the University of Minho and member of the C-TAC research centre. Lectures on undergraduate and postgraduate courses related to construction management.

Rania Wehbe (PhD)
Assistant Professor at Junia HEI–Lille Catholic University in France. PhD in civil engineering from Lille University in France. Interested in BIM & Digital Twins studies to improve the quality of life and safety of habitats in buildings and cities. Former president of the ASCE–USEK Branch.

Kitti Ajtayné Károlyfi (PhD)
Assistant Professor at Széchenyi István University in Győr. Teaching activities in the Department of Structural and Geotechnical Engineering, focusing on methodological questions related to BIM and interdisciplinary collaboration between architects and structural designers using BIM.

António Godinho (PhD)
President of ISLA-Gaia (Polytechnic University of Management and Technology) since 1999. Consultant in the computer systems area for several years. Founder of SISQUAL WFM.

Firmino Silva (PhD)
Adjunct Professor at ISLA-Gaia (Polytechnic University of Management and Technology). Head of the Research Unit at ISLA-Gaia, leader of the master's programme in Engineering of Technologies and Web Systems. Headed IT departments in various sectors.

Radhlinah Aulin (PhD)
An active lecturer and researcher attached to the Department of Construction Management, Lund University, Sweden.

Integrating Occupational Safety and Health using Digital Technologies

The case of Building Information Modelling

Manuel Tender, Paul Fuller, Peter Demian,
Matej Mihic, João Pedro Couto,
Rania Wehbe, Kitti Ajtayné Károlyfi,
António Godinho, Firmino Silva, and
Radhlinah Aulin

Routledge
Taylor & Francis Group

LONDON AND NEW YORK

Designed cover image: Shutterstock

First published 2026
by Routledge
4 Park Square, Milton Park, Abingdon, Oxon OX14 4RN

and by Routledge
605 Third Avenue, New York, NY 10158

Routledge is an imprint of the Taylor & Francis Group, an informa business

British Library Cataloguing-in-Publication Data
A catalogue record for this book is available from the British Library

ISBN: 978-1-041-01525-3 (hbk)
ISBN: 978-1-041-00399-1 (pbk)
ISBN: 978-1-003-61521-7 (ebk)

DOI: 10.1201/9781003615217

Typeset in Times New Roman
by Apex CoVantage, LLC

Contents

Background

The Digital4OSH Research & Development (D4OSH) team is based in PT and UK and have been collaborating since 2020. For this book we have brought together a team of academic and industry experts from across Europe. Their areas of expertise include Building Information Modelling (BIM), Occupational Safety and Health (OSH), and Information Technology (IT).

The book has emerged from a spirit of collaboration between academia, industry, and professional societies from different disciplines (e.g. Digitalization, OSH, Civil Engineering, Architecture, IT), with the aim of improving the sharing of knowledge across these areas to bring mutual benefits. The benefits of this approach include:

- Universities fulfil their mission by producing academic knowledge that can be supplied to the industrial community. It also allows universities to identify new research opportunities and their applicability to real-world production processes. It also increases their reputation and enhances their standing in the international community.
- Businesses can take advantage of highly specialized and technically advanced research facilities and infrastructures to operationalize their innovation activities, thereby refining their production processes, without the need to invest in their own research facilities. They can also use the created synergies to foster their own evolution, in terms of management and knowledge of markets, while they develop the skills of their human resources, accumulating physical and human capital. At the same time, they present themselves as sponsors of science to improve their reputation as socially responsible companies. Finally, since "the achievement of high levels of productivity will depend on the sophistication with which companies compete, the search for increasing levels of skills and technology is each time more important" [1]. This cooperation between academia and industry paves the way for improved productivity and an increase in the competitiveness of the companies involved in the project. By integrating the results of pure and applied research into the companies' production processes, the scientific knowledge created is put to practical use to improve the production process and corresponding financial consequences.

This collaboration will enable the book to present the synergies between academia and professional practice and promote the exchange and sharing of knowledge and experience between stakeholders, maximizing their strengths, breaking down barriers that may exist between them, and boosting sustainable and resilient evolution.

Improved knowledge transfer within the Digital4OSH community will enable the project to:

- Realize the synergies between academia and industry, maximizing their strengths, reducing barriers that may exist between them, sharing knowledge, boosting sustainable and resilient automation. This will capitalize on the fact that some of the countries involved have resources which have a high maturity and critical mass, arising from a high-quality scientific and business community.
- Create a transdisciplinary network bridging different research communities, disciplines, fields, and methodologies with people from different research fields, such as Digital Technologies, Computing, Information Systems, Civil Engineering, OSH, and Construction.
- Promote the exchange and sharing of knowledge and experience between experts from the research and practitioner communities ensuring the theoretical approach aligns with current practice;
- Make sure there is an efficient and worldwide dissemination and usage of the outputs of the research.

Follow us in

LinkedIn – https://www.linkedin.com/groups/9281223/
Facebook – https://www.facebook.com/groups/5335539779884970

About the authors

Authors

Manuel Tender (PhD)
Adjunct Professor at ISEP-Instituto Superior de Engenharia do Porto and at ISLA Polytechnic Institute of Management and Technology. Researcher at CONSTRUCT-Gequaltec, Faculty of Engineering, University of Porto. CDM Coordinator since 2001. Coordinator of the "Safety and Health" Working Group of Technical Commission 197 (BIM). Lead Researcher at the Digital4OSH Project.

Paul Fuller (ENGD)
Visiting Academic at the School of Architecture, Building and Construction Management of Loughborough University. Session Lecturer in Construction Project Management. 40 years of international experience in both large- and small-scale project environments.

Peter Demian (PhD)
MCIOB. Reader in Loughborough University. Professor of Digital Construction Engineering and Leader of Construction Management Group.

Matej Mihic (PhD)
Assistant Professor at the University of Zagreb Faculty of Civil Engineering in the Department for Construction Management, teaching courses related to construction safety, construction management, and construction technologies. President of Croatian Association of Health and Safety Coordinators in Construction.

Co-authors

João Pedro Couto (PhD)
Assistant Professor in Civil Engineering Department at the University of Minho and member of the C-TAC research centre. Lectures on undergraduate and postgraduate courses related to construction management.

Rania Wehbe (PhD)
Assistant Professor at Junia HEI–Lille Catholic University in France. PhD in civil engineering from Lille University in France. Interested in BIM & Digital Twins studies to improve the quality of life and safety of habitats in buildings and cities. Former president of the ASCE–USEK Branch.

Kitti Ajtayné Károlyfi (PhD)
Assistant Professor at Széchenyi István University in Győr. Teaching activities in the Department of Structural and Geotechnical Engineering, focusing on methodological questions related to BIM and interdisciplinary collaboration between architects and structural designers using BIM.

António Godinho (PhD)
President of ISLA-Gaia (Polytechnic University of Management and Technology) since 1999. Consultant at computer systems area for several years. Founder of SISQUAL WFM.

Firmino Silva (PhD)
Adjunct Professor at ISLA-Gaia (Polytechnic University of Management and Technology). Head of the Research Unit at ISLA-Gaia leader of master's programme in Engineering of Technologies and Web Systems. Headed IT departments in various sectors.

Radhlinah Aulin
Lecturer at Department of Construction Management, Lund University, Sweden. Leader for courses in Construction Management in Health and Safety and Project Management. Member of the international networking CIB W099 for Construction Health and Safety. Board member with CREON Home of Construction Researchers on Economics and Organisation in the Nordic Region.

Acknowledgements

Throughout our lives we have been fortunate and honoured to meet many people, from various walks of life, who inspired us, both personally and professionally. Many of them were, and still are, true beacons showing the right path and the right way. Saying that we came across hundreds of people, in Portugal, the United Kingdom, and other countries, would be an understatement. Natural affinities grew into professional partnerships, and some of those grew into friendships. To all those who, in one way or another, collaborated in this research, served as a reference, provided guidance, or "gave us a hand" when things were hard or helped us overcome the problems we faced, we leave here some words of thanks, sparse though they may be:

- To our families, fathers and mothers, spouses, children, grandchildren;
- To the Deans, Directors, and colleagues of our Schools of Engineering;
- To our students and researchers;
- To the companies which supported the Research and Development work linked to this book:

Digital4OSH founders:

> ISLA-Polytechnic Institute of Management and Technology, School of Technology Loughborough University
> Glasgow Caledonian University
> School of Engineering of Minho University
> BIMMS Management
> Xispoli Engineering

Institutional partners:

> BuildingSmart Portugal
> CONSTRUCT – Gequaltec, Faculty of Engineering, University of Porto
> ISEP, Instituto Superior de Engenharia do Porto
> Grupo Académico ISEPBIM
> Lund University

Junia HEI – Université Catholique de Lille
Faculty of Civil Engineering University of Zagreb
Széchenyi István University

Technical contents suppliers

Metro do Porto
Ferrovial Construccion
Alberto Couto Alves
Electrofer
PERI
Fuzor
Solibri
BIMSafe

This work was financially supported by Base Funding – UIDB/04708/2020 with DOI 10.54499/UIDB/04708/2020 (https://doi.org/10.54499/UIDB/04708/2020) and Programmatic Funding – UIDP/04708/2020 with DOI 10.54499/UIDP/04708/2020 (https://doi.org/10.54499/UIDP/04708/2020) of the CONSTRUCT – Instituto de I&D em Estruturas e Construções – funded by national funds through the FCT/MCTES (PIDDAC).

Abbreviations

AECO	Architecture, Engineering, Construction, and Operations
BCF	BIM Collaboration Format
BEP	BIM Execution Plan
BIM	Building Information Modelling
CDE	Common Data Environment
CEN	European Committee for Standardization
EIR	Exchange Information Requirements
EU	European Union
IFC	Industry Foundation Classes
OSH	Occupational Safety and Health
O&M	Operation and Maintenance
PAS	Publicly Available Specifications
SME	Small- and Medium-Sized Enterprises

Foreword

The aim of this book is to provide a guide to both academic researchers and practitioners on how to improve the results of OSH actions aimed at zero accidents and a healthier working environment. For no other reason the book recommends itself by wishing for a better future for all. Another reason for accessing this book is due to the backgrounds of the authors and co-author who are experienced professionals and active researchers in OSH who can provide a richness of case studies and lessons learned. A further reason is the knowledge provided about the new tools and digital techniques being used for OSH in construction. It helps overcome the conservative attitudes towards innovation by presenting real-world examples sector to the 21st century. Finally, the book provides guidance and actions that can help stakeholders develop action plans. A key area is the training supported by digital tools that can bring better ways to prepare technicians, workers, suppliers, students, officials, and OSH practitioners to handle difficult and dangerous situations after acquiring competences related with OSH.

The adoption of BIM has evolved with the change from just modelling to management of information which provides data that can be used to simulate and to assess the different scenarios throughout the construction life cycle. The issue is that each construction site is unique and requires specific solutions in terms of OSH. This means that the idea of the European directive 92/57 which tried to implement the notion that OSH in construction should be treated as any other design element, like reinforced concrete or Heating Ventilating Air Conditioning (HVAC). Data management using BIM provides the possibility of simulating and of optimizing the OSH preventive measures within the constraints of limited resources that are generally available from construction companies. BIM tools can combine data from previous accidents related to each one of the tasks to be executed, evaluate the risks, and use the financial resources to decrease the impact despite the fact that it is impossible to eliminate all risks in the construction sites. It will be easier and more effective using these digital tools to have a more efficient way of using the time of OSH professionals and of distributing the safety equipment to achieve a better outcome. If this book saves at least the life of one worker that is a sufficient reason to write it and to publish it. Thank you for publishing the book!

Professor Alfredo Soeiro
Faculty of Engineering of University of Porto

Foreword

In an industry as dynamic and impactful as the Architecture, Engineering, Construction, and Operations (AECO) sector, the integration of innovative methodologies and technologies is no longer an option – it is an imperative. Among these innovations, Building Information Modelling (BIM) stands out as a transformative approach, reshaping how we design, build, and operate our built environment. This book, *Integrating Occupational Safety and Health using Digital Technologies: The Case of Building Information Modelling*, represents a milestone in leveraging BIM to address one of our industry's most pressing concerns: ensuring the safety and well-being of everyone involved.

The vision of BuildingSMART has always been to promote openBIM standards, fostering interoperability and collaboration. These principles are aligned with the themes of this book, which highlights how digital technologies can deliver safer, more efficient, and more sustainable construction practices. By integrating Occupational Health and Safety (OSH) considerations directly into BIM processes, this book not only demonstrates the potential for accident prevention but also showcases how the AECO sector can advance in terms of maturity and responsibility.

It is particularly gratifying to see such a well-researched and practical work emerging from collaborations between experts globally, with contributions from academia and industry. The insights shared here will undoubtedly inspire practitioners, researchers, and policymakers to rethink their approach to safety and health management in construction projects.

For all this, I congratulate Manuel Tender and all the authors for their dedication and expertise in producing this timely and important work. I encourage all readers to embrace the opportunities presented in these pages, and the transformative power of digital technologies in our shared mission to build a better future.

Let's commit to a safer, smarter, and more collaborative AECO sector.

Eng. José Carlos Lino
President, BuildingSMART Portugal

Chapter 1

Introduction

1.1 Overview

Digital Technologies in the Architecture, Engineering, Construction, and Operation (AECO) sector have created a step change in the traditional ways of dealing with Occupational Safety and Health (OSH) risk management during the life cycle of buildings, enabling digitalization to minimize risks and costs of accidents at work (AW) and occupational diseases (OD). One important area where digital technologies have so far not played a significant role is OSH. At the same time and in several cases, the current way of managing OSH often takes the form of a long list of procedures which are not always understood by those who have to implement them. In many cases the procedures are not properly analysed by those responsible for supervising the performance of tasks, and there is a lack of integration with the planning of the work. This leads to OSH taking a secondary role and not being sufficiently valued. Therefore, the number of AW remains far too high, with a corresponding financial and logistical impact for companies. There has been little research into the use of new technologies for OSH in real practical cases in Europe.

This handbook focuses on the potential of one such Digital Technology, Building Information Modelling (BIM), to address OSH problems and improve OSH outcomes. The purpose of this handbook is to address the different ways of integrating preventive information in the different dimensions of the BIM model, in terms of both risk identification and preventive measures, namely the areas of document and contractual management, risk identification, training, site and task planning and monitoring, emergency planning, and investigation of AW.

This will raise the bar and encourage the wider introduction of BIM for OSH as a strategic enabler adopting an aligned framework for its introduction into the AECO sector. This alignment enables clarity and repeatability to this digital innovation reducing divergence, misunderstanding, and waste; accelerating growth; and encouraging competitiveness of the construction sector, especially its small and medium enterprises (SMEs) and paving the way for better streamlining of prevention planning, by making it more visible and understandable. It also allows for OSH to be closely linked to production, thus creating the conditions for an effective integration of OSH within the processes.

DOI: 10.1201/9781003615217-1

This approach has the potential to leave an important legacy, breaking through the current boundaries of knowledge capture and transfer in both academia and professional practice.

This is a first effort not only in construction OSH but also in the construction management research, so that the general principles of this book can be replicated for future research of other technologies, for example, Internet of Things and Digital Twins.

The objectives of this book are to

- analyse the potential of integrating OSH in digital data using BIM;
- identify the enablers, difficulties, and the obstacles to implementation;
- identify the lessons learned from existing approaches to integrate OSH in digital data using BIM, to improve OSH outcomes in industry practice;
- improve the skills of the research community and encourage the development of new research;
- minimize risks and costs related to AW and OD;
- develop awareness and encourage the use of digital technologies to improve OSH outcomes; and
- identify the new key skills needed for OSH workers.

Providing construction stakeholders with a list of areas where BIM is used for OSH management and identifying their benefits, limitations, and barriers will help improve the stakeholders' knowledge and understanding of BIM.

Providing construction stakeholders with a description of how to integrate OSH into digital data using BIM and identifying the main benefits, limitations, and barriers of such an endeavour will boost knowledge and understanding about BIM. This, in turn, will encourage greater adoption of similar solutions, especially among actors who are less determined or more resistant to change. In this way, safety performance across the industry can be improved.

This book has a large potential target audience, and we expect it to be of interest to both the industry community (large companies and SMEs, project owners (POs), designers, safety coordinators in the design and in the construction phases, contractors, OSH and BIM technicians, inspectors, construction managers, consultants, suppliers, standardization or regulatory agents, software and hardware providers, societies, agencies, associations and confederations, policymakers, and technicians, namely in infrastructure projects, rehabilitation, and housing) and the academy community (researchers, teachers, and students).

The book is divided into strongly interconnected chapters. Chapter 1 outlines the work and sets it in the context of current situation practice, demonstrating its relevance. It defines the problem and its significance, the objectives, the organization of the research questions, the target audience, the organization of the bibliographic review, and the survey of the state of the art. It also focuses on the actual framework of risk management and BIM. The next chapters are dedicated to specific uses of BIM: documental management, risk identification, training, site planning, tasks planning, monitoring, emergency planning, work accident investigation, and operation and maintenance (O&M). Chapter 10 explores technologies beyond BIM. Chapter 11 describes challenges and barriers to BIM adoption in OSH. Chapter 12 draws conclusions, the practical implications, and the future trends. Chapters 13 lists useful websites, training courses, and software suppliers in BIM area.

1.2 AECO sector

As of 2024, the AECO sector remains a pivotal contributor to the global economy, marked by its expansive scale, impact on other economic sectors, and evolving challenges. This sector is notably diverse, encompassing activities from architectural design and engineering to construction and long-term building management. According to a report by The Business Research Company, the global construction market is expected to grow from $12.7 trillion in 2020 to over $15 trillion by 2025 (The Business Research Company, 2021). Economically, the AECO sector is a significant employment driver, with Eurostat reporting that construction alone accounts for approximately 6% of total EU employment (Eurostat, 2023). The sector's economic contribution is not just limited to direct construction activities but also extends to related fields such as manufacturing of construction products, design services, and facility management. Despite the production drop caused by the global COVID pandemic, and the effect of more recent international conflicts, the production is expected to continue along its growing trajectory.

However, the sector faces numerous challenges, including sustainability concerns and labour shortages. The pressing issue of climate change has put the AECO sector at the forefront of sustainability debates, with an increasing focus on green building practices and energy-efficient designs. The European Construction Sector Observatory (ECSO) highlights the industry's shift towards sustainable construction and digitalization as key trends (Observatory', E. C. S. 2023).

As of 2023, the AECO sector is in a state of dynamic evolution, adapting to new technological advancements and societal demands, particularly in the areas of sustainability and digital integration.

1.3 OSH management

Construction is one of the sectors with the highest rate of fatal and non-fatal AW among all economic activities. This can be understood as the methodologies currently used are not very effective at reducing the risks and the number of occupational accidents in construction.

Risk can be considered the effect of uncertainties in achieving objectives and can be characterized by reference to potential events and consequences or a combination of both. Risks are threats to the fulfilment of predetermined objectives and can assume a positive or negative character (Taylan et al., 2014). In this study, in line with the vast majority of studies in this area, only the negative aspects of the risks' impact will be addressed. There are several types of risk: political (Gafari and Aminzadeh, 2015); legal, namely legislative non-compliance (Gafari and Aminzadeh, 2015); contractual and financial, for example, insolvency and institutional problems (Lena and Eskesen, 2006), non-compliance with deadlines and estimated budgets; technical, for example, inadequate design, specifications or planning (Lena and Eskesen, 2006), adverse conditions (Lena and Eskesen, 2006), exposure to construction techniques, equipment, and materials introduced by technological developments (Longo, 2006); work accidents and OD (Tender et al., 2016); risks for workers and third parties (Lena and Eskesen, 2006); natural (Longo, 2006), for example, floods, landslides, falling blocks, hurricanes, lightning, infestations, earthquakes, volcanic eruptions; and anthropological (Longo, 2006), for example, the type of risk due to pollution and environmental degradation of flora and fauna (Lena and Eskesen, 2006), caused by human activities, responsible for the existence of events such as acid rain, contamination of surface and underground water, ozone depletion, and greenhouse effect. In this book, only technical risks will be analysed.

Currently, risk management emerges as an effective procedure that complements the management of almost all aspects of human life (Marhavilas et al., 2011). Risk management is a process that aims to identify the dangers and risks associated with each job, estimate the magnitude of the risks that cannot be avoided, compare it with reference standards to establish the degree of risk acceptability/tolerability, determine the most appropriate preventive measures to minimize these risks, and define procedures for monitoring and reviewing risks. It is assumed, however, that it is impossible to eliminate all risks (KarimiAzari et al., 2011).

The need for risk management arises in four ways: (a) by legal imposition; (b) to comply with the general principles of prevention; and (c) by the influence it has, in terms of schedules and costs, on the occurrence of AW and OD. Each of these needs can be addressed in more detail:

a) Legally, risk management is a theme referred to in most of the legal diplomas related to prevention (Carvalho and Melo, 2013), since it is the employer's obligation to ensure the integration of the risk assessment for the safety and health of the worker with the company activities. Normalization is applicable to OSH management through the International Organization for Standardization (ISO) 31000: 2018 (Leitch, 2010), supporting a four-step approach to risk: planning, implementing, monitoring, and reviewing.

b) With regard to the general principles of prevention described in Directive 89/391/EEC – European Framework Directive (European Commission, 1989), the central role that risk management plays in almost all 11 principles is evident.

c) The occurrence of AW and OD could necessitate work to stop for hours, days, or even weeks, in a given work front, with consequent and high economic and social implications (Hermanus, 2007), resulting from direct and/or indirect costs (productivity loss, absenteeism, low morale, loss of yield, compensations for damages, time spent in the analysis of the event). This suspension of works may be a short one, with the front released within hours. But it may also be a long one, pending the collection of all the necessary data for the investigation of the accident. Although accidents may not traditionally have been considered as one of the main causes for delays or non-compliance with deadlines, it appears the occurrence of accidents can be decisive for the conclusion of the work. In some cases, it may even compromise the success of the project (Couto, 2007). So, accidents can have a relevant impact on the term of the work, which should certainly catch the attention of company managers, and should always be considered in the analysis of this type of problem (Shannon et al., 1999). In addition, there is the issue of associated costs. Stoppages could have high economic and social implications, with consequent direct and/or indirect costs (loss of productivity, absenteeism, low morale, a decrease in productive income, compensation costs, and time spent analysing the accident). These costs would certainly affect, through a decrease in the company's profit margins, the company's competitiveness, as well as its financial results. O&M costs are generally neglected during the design and construction phase, despite evidence to suggest that they contribute to more than half of the total life cycle costs (Heaton et al., 2019). Adding to that, the company would be less likely to become a preferred supplier, namely for POs where prevention is among the first concerns.

The implementation of a risk management system has the following advantages (Haslam et al., 2005):

- Early identification of potential hazards and tasks at greatest risk;
- Quantification of risks and comparison to acceptable and tolerable risk criteria, thus establishing priorities in their treatment (Carvalho and Melo, 2013);
- Timely decision-making on the preventive actions most appropriate to the risks identified;
- Monitoring of residual risks;
- Elimination of the increase in costs associated with work stoppages or the drop in income derived from the occurrence of AW (Longo, 2006);
- Helping decision-makers to allocate resources (time, money, equipment, workers) to manage the most critical risks (Fine, 1971).

It is now assumed that risk management is part of the decision-making process (Mahdevari et al., 2014) and the foundation for a successful and proactive Health and Safety system (Ceyhan, 2012), the basis for effective safety management, and the key to lowering AT and DP (Lena and Eskesen, 2006).

Risk management must be present during all stages of the project. Right at the onset, during the design phase, managing risk by establishing a risk policy and risk acceptance criteria influences important decisions, such as the architectural layout or the constructive process. Risk management is also important through the tender and negotiation phases because it influences, among other things, the requirements in the tender, the risk assessment in the tender phase, and the integration of risk clauses in the contract. When the enterprise reaches the construction phase, risk management should be carried out jointly by the contractor and the work owner.

The risk management process is complex and time-consuming due to the diversity of tasks to be considered for analysis and the complexity of the dangers and risks present in each of them.

The risk management of occupational hazards is influenced by many factors including space components, hazard characteristics, its vulnerability, and worker exposure to hazard. Therefore, the understanding of the hazard, its causes, and consequences help in providing a better monitoring system.

The risk management process itself is structured in several phases (Carvalho and Melo, 2013):

- Identification of hazards, exposed people, and possible consequences;
- Identification of risk;
- Estimation of risk associated with hazards (these three phases form the risk analysis, which aims to determine the magnitude of risk);
- Valuation of risk;
- Assessment of the importance of a given risk, by comparing the level of risk that was obtained with the level of acceptable or tolerable risk (obtained in the phases prior to the risk assessment);
- Risk control;
- Risk reduction measures and mitigation measures.

There are several risk control strategies that can be progressively adopted, depending on the importance of the risk and the cost/benefit ratio:

- Eliminating the risk at its source by choosing a different constructive technique, which is the usual (and better) principle of any risk management programme (Mahdevari et al., 2014);
- Avoiding the risk, by deciding not to start or continue with the activity that gives rise to it;
- Reducing the risk through reduction of probability and/or consequence (Goh and Abdul-Rahman, 2013) by improvements in construction processes or with preventive measures (Mahdevari et al., 2014) to bring the risk down to acceptable levels. These preventive measures may be of a technical, administrative, organizational, collective, or individual nature;
- Accepting the risk and its level without trying to control it;
- Sharing the risk (Goh and Abdul-Rahman, 2013) through public-private partnerships or contractual relationships with suppliers;
- Transferring the risk (Goh and Abdul-Rahman, 2013) by passing its effects to another entity (e.g. insurance company) without changing the risk level.

The position of the PO is of particular importance as it is their needs that ultimately drive the project. They will establish the environment in which the work will be carried out by contractors in the different phases. They may, therefore, directly influence the resources that contractors will allocate to prevention. Consequently, it is important that the PO understands the importance and influence they have in decisions related to preventive matters. It is essential that the PO has on-site personnel representing the PO, with the necessary civic and educational training for this purpose. This is also valid for all other stakeholders, but the PO will also serve as a guide for the desired tolerance level for the work.

The act of integrating prevention in the design phase using the documents made available in the construction and O&M phase must stem from the PO's actions. The fact is that this integration can bring several gains, be it in technical, logistical, or financial terms. It allows time to identify potential deviations and deal with them before it is too late, and the risk becomes unacceptable. The safety of any operation is determined long before workers and equipment come together at the work site (Behm, 2005). There are some risks that still materialize on site which are those that have been impossible to eliminate during the design phase. However, even those would be better managed as the designer would have informed the contractor of their existence which would then make it easier for the PO to manage them.

To ensure a correct risk management, the first fundamental step is reliable identification of hazards (Badri et al., 2013), that is, the sources of potential damage from risk conditions or unsafe acts which occur in the work system that may contribute to the occurrence of AW and OD (Khanzode et al., 2011). This step can be considered as the most critical of the whole process in that an unidentified hazard may not be evaluated and consequently, uncontrolled (Carvalho and Melo, 2013) which gives rise to risks that cannot be managed. Otherwise, it would not be possible to recover from errors in later phases with risk management assuming a false structure (Ceyhan, 2012). In this initial, essential phase, the basis for risk analysis is established, to assess and identify the dangers present in each of the work phases and those associated with substances, equipment, procedures, environment, dangerous conditions, or unsafe acts, as well as the corresponding consequences in terms of damage that may be suffered by workers exposed to them. It is important to include here all activities/tasks, be they routine, occasional, or emergency. For this phase, one should be able to answer the following questions, for each task: "What can go wrong?"; "How likely is that to happen?"; and "If that happens, what are the likely consequences?" [12]. With the answer to these questions, the identification of hazards associated with each task will be accomplished. It is, therefore, crucial to establish the correct way to ensure consistency and correction of hazard identification (Ceyhan, 2012). It should be noted that the analysis of risks in construction is not necessarily based on logical reasoning, but on predictions guided by experience (Mordue and Finch, 2019). Therefore, the individual assessment of each risk is very important.

The identification of risks must be based on the most complete, detailed, and reliable data possible (Labagnara et al., 2013), and we must not forget that all projects are different, and they all have risks that can be unique.

From the above, it is clear that the ideal way to approach safety proactively is to strive from the beginning of the project's development (Behm, 2005) to identify and eliminate or, if this is not possible, to minimize the risks that construction activities carry. The safety level is defined by a set of choices that have, among others, prevention as a criterion that integrates the 11 general principles of prevention (Tender et al., 2015). This is translated into the selection of construction materials and definition of the standards required for the construction and O&M phases.

One way of reducing risks is the establishment of adequate and preventive risk reduction measures, to be implemented by the employer, for the protection of workers' safety and health (European Commission, 2008). The establishment of preventive measures, in addition to helping to minimize risk, has several advantages, for example, defining responsibilities, allowing cost allocation, helping with emergency planning. These measures can take different forms (presented below in order of importance; Ceyhan, 2012):

Organizational – related to construction processes, site organization. They require a change in the way the activity is performed.
Collective – isolate workers from danger, through technical and administrative procedures.
Individuals – as the name implies, such measures act on the individual and should be the last resort in preventing accidents. It should be noted that there is often a need to combine different types of preventive measures.

(Ceyhan, 2012)

OSH process outputs are typically visual, for example, maps or reports that present the results of the risk assessment done and are understandable by all involved. Through these outputs, risks are identified and classified and preventive measures are established.

Construction remains a high-risk activity, with several types of risk, including physical (e.g. noise, dust, vibration), chemical, ergonomic (e.g. heavy loads, uncomfortable positions), and psychosocial (e.g. stress) risks. Construction work is also known as dirty, tough and hazardous, highly manual, and transient in nature.

In fact, in spite of developments in construction and OSH processes that may have affected the AW and OD rates in previous years, construction is still one of the most dangerous fields of work (Godfaurd and Abdulkadir, 2011), with a large number of AW and OD keeping accident rates quite high (Azhar and Behringer, 2013), both during construction and in the O&M phase. These accident rates have a considerable financial and logistical impact on companies, with short-, medium-, and long-term repercussions in reputation and finance (Zou et al., 2017).

Accidents are unfortunately not a rare phenomenon in the construction industry, with it being one of the most hazardous industries. And while an accident is an incident which causes harm to a worker, for every accident there are an even greater number of near misses, which are incidents that have not caused harmful effects but could have, if circumstances were a bit different. But what can cause an accident? Through the years there have been numerous accident causation theories. First starting from theories that exclusively a person's traits and their unsafe behaviour are responsible for accidents, then moving to more complex causation such as Heinrich's Domino theory, which states that certain chain of events needs to occur in order for an accident to manifest. The name implies that if one domino is removed, the chain will be broken and not all the dominoes will fall. In concept, this is similar to the Swiss cheese model by Reason which posits that in order for accident to occur, multiple gaps in layers of defence (holes in the cheese) need to align. This analogy holds true in practice because no single measure is perfect, nor does simply an existence of an error lead to an accident automatically occurring.

Khanzode et al. (2012) has divided accident theories in four generations, first two of which were covered earlier and which place the causation on unsafe acts and on both unsafe acts and unsafe conditions. The third generation includes environmental factors, while the fourth generation takes an integrated systems approach to explain accident causation (Khanzode et al., 2012). Accident causation is indeed complex to determine since there may be underlying factors which are not evident at first, due to the fact that immediate or primary causes of accidents, such as unsafe acts and unsafe conditions, are often made possible by the underlying secondary causes which allow the primary causes to appear and to exist, by not removing them. From this short excursion into accident causation, it can be seen that in order for accidents to be avoided, there is a need to place as many barriers as possible. These would safeguard against human errors, equipment failures, and environmental factors. The use of digital tools can aid in adding additional layers and strengthening existing ones.

Despite its clear usefulness, the risk management process is not always viewed in a positive light and as an opportunity (Pasman, 2015) or even as something more than the mere need to comply with legal requirements (Rodrigues et al., 2016). This is perhaps due to cultural or financial issues. It is often seen as reactive (e.g. in the event of imminent danger) rather than preventive. Research also suggests that the construction industry is saturated with traditional injury-prevention strategies. Naturally, a crisis scenario (which translates into tight execution deadlines, high production rates, budgets with low financial margins, and very aggressive competition) aggravates this way of thinking about prevention which makes implementation difficult. However, sometimes, risk management does not reach its ends being little more than a documentary, static, and administrative exercise that distorts the spirit of true risk approach. All these tend to increase as the size and complexity of projects increase, despite improvements in construction processes. It should be noted that as companies seek to make profits, not only in the design and construction phases but also in O&M phase. However, O&M costs tend to be neglected during the design and construction phases, despite evidence suggesting that they contribute to more than half of the total life cycle costs (Heaton et al., 2019).

In reality, except for the larger projects, the implementation of the OSH management is far below what would be desirable, and it is necessary to create mechanisms to fill the identified gaps and take risks down to acceptable levels. This whole scenario brings about an environment where these issues are downgraded in terms of both risk identification and preventive measures, thus giving prevention, a secondary role not integrated in the execution of the work or in the maintenance of the building (Tender et al., 2018c). This scenario also leads to a higher risk of AW and OD (Haslam et al., 2005) and tends to undermine the challenge of integrating safety in the planning of works. As safety should always be of prime importance, regardless of deadlines and economic interests, a change of mind-set is needed (Tender et al., 2015).

1.4   Digital technologies

Society today is undergoing a new Industrial Revolution, often referred to as the fourth Industrial Revolution. The first Industrial Revolution saw the transition from hand-based production to machine-based production; the second Industrial Revolution came with the industrialization of production processes; the third one was related to the automation of processes, with computers handling large amounts of information, minimizing calculation times, and optimizing the quality of products; and the fourth, currently underway, is related to the digitalization of production processes. This fourth Industrial Revolution is changing the way data is shared, and stakeholders cooperate within the production process. The era of digitalization of information encompasses technologies for automation, through sharing and exchanging data, and the use of virtual graphic models. It has been accelerated by the ability to access documents via smartphones and wireless networks. Additionally, as the use of social networks increases, so does collaboration between all actors which optimizes the spread of information (Mordue and Finch, 2019).

This metamorphosis is being driven by the rapid advancement and adoption of emerging information technologies. These technologies are not only changing the way construction projects are executed but also reshaping the entire ecosystem of the construction sector, enabling them to be the steppingstone for a new era of complexity and connectivity within the industry. Traditionally, construction has been associated with manual labour, blueprint drawings, and on-site decision-making. However, in recent years, a wave of technological innovation has swept across the industry, revolutionizing age-old practices and introducing unprecedented levels of efficiency, safety, and sustainability. We are now witnessing the advent of smart construction sites, data-driven decision-making, and the fusion of digital and physical elements in the built environment. Digitalization of the AECO sector is increasingly being recognized as a potential game-changer for the sector, enabling an economy that grows based on sustainability, smart knowledge, and innovation. By applying such digital technologies and using smart and interconnected cyber-physical systems, people, machines, equipment, logistics systems, and products communicate and cooperate directly with one another during all life cycles of project. The construction industry, often characterized by its traditional and resistant-to-change nature and slow pace of adoption of innovative approaches, is currently undergoing a profound transformation. However, growth rate in this sector is much lower than that in other sectors. This is partly explained by the difficulties of the building sector in embracing digital innovations that could help improve both productivity and profitability. There is also still a noticeable gap between large construction companies and SMEs, with SMEs having a much lower uptake of digital tools.

The industry is moving towards an integrated digital approach, which is not only reshaping the way projects are delivered but also promising improved sustainability and efficiency of built environment assets (Fernández-Solís et al., 2020).

It is estimated that full-scale digitalization in non-residential construction would lead to annual global cost savings of EUR 0.6 trillion to EUR 1.0 trillion (13%–21%) in the engineering and construction phases and EUR 0.3 trillion to EUR 0.4 trillion (10%–17%) in the operations phase (European Commission, 2019).

We will delve into some new technologies that are the basis of a new approach to the AECO sector and are spearheading this transformation: BIM, Artificial Intelligence (AI), Wearables, Big Data and Internet of Things (IoT), Virtual Reality (VR) and Augmented Reality (AR), 5G connectivity, and Drones. The application of each of those components to the design, construction, and operation of the built environment enables fostering better communication, reducing errors, and improving overall project efficiency. These technologies offer construction professionals the ability to work more effectively and collaboratively, ultimately leading to projects that are completed on time, within budget, and with enhanced safety and quality. They also can have significant implications for effective OSH management. For each of these developments, we will explore their unique advantages, their disruptive potential to upend traditional construction approaches, and the practical applications that are making construction safer, more efficient, and environmentally sustainable. It also recognizes that while these technological advancements bring tremendous opportunities, they are not without their challenges. Concerns regarding data security, workforce readiness, and the need for adaptation to these rapidly evolving technologies must be addressed. Yet, as we will see, the advantages and potential for innovation far outweigh the obstacles, offering stakeholders in the construction industry the prospect of not just surviving but thriving in this era of technological disruption. All these new technologies are different but complementary to each other and can be used in an integrated way to enhance workers' real-time communication ability in an ever-changing environment.

Building Information Modelling (BIM): It is an intelligent, three-dimensional (3D) model–based toolset that provides a digital representation of a facility's physical and functional aspects, a digital representation of the physical and functional characteristics of a building or infrastructure. Like an Enterprise Resource Planning (ERP) system, it serves as a robust instrument that consolidates various data origins and extends its utility beyond the realm of design and construction, offering the capability to elevate multiple facets throughout a building's lifespan. BIM integrates functions that make it an innovative and extremely useful tool: it also allows architects and engineers to create detailed 3D models of a building before it is constructed.

Artificial Intelligence (AI) and Machine Learning: the application of AI algorithms and learning mechanisms (Machine Learning [ML]) enables a new disruptive approach. They furnish anticipatory insights, empowering on-the-fly decision-making, risk management, and diminishing uncertainties in project execution. Supported by ML algorithms, they optimize construction schedules and resource allocation. In parallel, AI is increasingly being applied to predict outcomes, enabling predictive maintenance, energy optimization, and enhanced design decision-making, personalize experiences, and automate tasks that traditionally required significant human labour. The capability of AI to analyse and learn from data can dramatically improve the accuracy and efficiency of various AECO tasks (Oesterreich and Teuteberg, 2016).

Wearables, IoT, and Big Data: Wearable technology, IoT, and Big Data analytics are increasingly making their presence felt in the construction industry, ushering in a new era of digitization and efficiency. These devices are capable of (in real time) tracking vital signs, detecting fatigue, and even monitoring environmental factors like temperature and air quality. Integrating the IoT has propelled the AECO industry towards smarter asset management and operational practices, enabling the collection of extensive data from various sensors and providing real-time insights into building performance, thereby facilitating proactive maintenance and energy management (Lu et al., 2017). This integration is critical for advancing intelligent building concepts that autonomously optimize for sustainability and occupant comfort. Construction materials embedded with IoT sensors can provide real-time information on their condition, ensuring quality control and reducing material wastage. Additionally, IoT facilitates remote monitoring and management of construction sites, allowing project stakeholders to oversee operations from anywhere, enhancing collaboration, and improving overall project efficiency. As the construction industry continues to embrace these technologies, the potential for enhanced safety, productivity, and sustainability becomes increasingly promising.

Virtual Reality (VR) and Augmented Reality (AR): VR, on the one hand, immerses users in a computer-generated environment, allowing them to explore and interact with 3D models of construction sites, buildings, and infrastructure. It is particularly valuable during the design and planning stages. Architects and engineers can step into a virtual construction site, visualizing the project from various perspectives and making real-time design modifications. This immersive experience helps identify potential issues, enhancing design accuracy and reducing costly revisions during the construction phase. AR, on the other hand, overlays digital information onto the real-world environment. AR and VR experiences become seamless and lag-free, providing stakeholders with the ability to visualize construction plans, explore 3D models, and simulate on-site conditions in real time. This not only enhances collaboration among design and construction teams but also enables more accurate design validation and training exercises, ultimately leading to safer and more efficiently executed projects. Construction professionals wear AR headsets or use mobile devices equipped with AR applications to superimpose 3D models, plans, and data onto physical construction sites. AR enhances on-site decision-making by providing real-time access to critical information, such as blueprints, project schedules, and safety protocols. Workers can visualize where utilities are located beneath the ground or identify structural components within walls, improving accuracy and safety. Moreover, AR facilitates remote collaboration, allowing experts to provide guidance and instructions to on-site personnel in real time, even when they are miles away.

5G Connectivity: the rollout of 5G connectivity is poised to be a game-changer for the construction industry, offering a quantum leap in communication and data transfer capabilities. With its ultra-fast data transfer speeds, low latency, and network reliability, 5G opens a plethora of possibilities in construction project management. One of the most immediate impacts is in remote construction management. Project managers can access construction site data, monitor progress, and communicate with teams in real time from virtually anywhere in the world. This real-time access to critical project information enhances decision-making, expedites issue resolution, and reduces the need for physical presence on-site, which can be particularly beneficial in remote or large-scale projects. These immersive technologies can revolutionize design reviews, training, and remote inspections. Furthermore, 5G connectivity enhances the capabilities of IoT devices that benefit from the higher data transfer rates and low latency of 5G networks. This results in more real-time monitoring and data analysis. Overall, 5G connectivity promises to be a transformative force in the construction industry, empowering stakeholders with faster and more reliable communication, data sharing, and innovative applications that drive efficiency, safety, and collaboration.

UAV (Unmanned Aerial Vehicle) and UGV (Unmanned Ground Vehicle): UAVs and UGVs have rapidly become indispensable tools in the construction industry. They offer a unique perspective from the sky, providing valuable data and insights that were previously inaccessible or time-consuming to obtain. If equipped with high-resolution cameras and sensors, they can capture detailed aerial imagery, create 3D maps, and monitor construction sites in real time. This data is instrumental in site planning, project management, and quality control. One of the primary advantages of these equipment in construction is their ability to perform aerial surveys and site inspections quickly and cost-effectively. Traditional methods, such as manned aircraft or ground surveys, are often expensive and time-consuming and may pose safety risks. This real-time data aids in decision-making, allowing project managers to respond promptly to changes or discrepancies.

1.5   Building Information Modelling (BIM)

1.5.1 Defining BIM

BIM is commonly understood to stand for Building Information Modelling (Barnes and Davies, 2014). However, alternative terms have been proposed, including phrases such as Building Information Management (Barnes and Davies, 2014),

Building Information Models, and Better Information Management (NBS, 2021). Barnes and Davies even propose combining phrases to create the acronym BIM(M) to stand for "Building Information Modelling and Management" (Barnes and Davies, 2014).

There is little consensus in defining BIM. Ahmad et al. (2012) counted the instances when various concepts (Information, Analysis, Process, Management, Modelling, Technology, and Collaboration) were invoked when defining BIM and reported a wide range of concepts, leading to many inconsistent definitions (Ahmad et al., 2012). Up to the time of the UK Government BIM mandate, definitions seemed to focus on representing 3D constructed facilities to aid decision-making during the processes of designing, constructing, and operating them. For example, in 2011, the BIM working party defined BIM as the "digital representation of physical and functional characteristics of a facility creating a shared knowledge resource for information about it forming a reliable basis for decisions during its life cycle, from earliest conception to demolition" (Rodrigues et al., 2016). In more recent definitions, the emphasis shifts from representing products to managing information. For example, the Centre for Digital Built Britain (CDBB) writes that BIM

is a set of digital tools, processes and standards for information management used to capture and store the data associated with a construction project so that it can be shared by everyone working on the build and those responsible for the assets' subsequent operation.

(CDBB), 2018)

In fact, BIM is a collaborative methodology based in a 3D model with parametrized information that provides a digital representation and remotely accessible by all intervenient (PO, the designer, the contractors, the suppliers, the consultants, etc.) of the physical and functional characteristics of a building or infrastructure encompassing the whole life cycle of the project, from design to construction and to O&M.

Throughout the research literature, the theme repeatedly emerges that BIM is not limited to a particular software platform and that BIM is not merely a noun. Rather, BIM is a also a verb and a collaborative process, a method of working, or a human activity (Tender et al., 2018c). BIM indeed originates from 3D Computer-Aided Design (CAD) models. However, the difference is that in BIM, 3D geometry is enriched with non-geometrical data and information which allow better, more collaborative decision-making throughout the building life cycle (Yan and Demian, 2008).

The concept of modelling in construction is not new, but quite the opposite. For many years, cardboard or wood models were used to represent the end result of the project (Mordue and Finch, 2019).

BIM has evolved from, and is rooted in, the concept of CAD. Depending on the literature considered, CAD started to emerge at some point between the 1950s and the 1980s (Barnes and Davies, 2014). CAD usage increased throughout these decades with the increasing availability and decreasing cost of personal computers. However, at this point, the majority of the software was developed in-house for design companies, with very few off-the-shelf packages available for general purchase (Barnes and Davies, 2014). Many point to Sketchpad as the first readily available CAD software. This was released by Ivan Sutherland in 1972 (Sardroud et al., 2018). However, others argue that the biggest turning point came in the next decade with Graphisoft's development of ArchiCAD in Hungary and Autodesk's AutoCAD in the United States (Yan and Demian, 2008). CAD, as the digitalization of drafting, was an important forerunner to BIM, which evolved from 3D CAD. Of course, nowadays, computers are used to produce the models using tools based in object-oriented modelling. These objects (constructive elements) include technical data including geometry, volume, density, spatial relationships, and data from suppliers which is used in the model based on a set of pre-defined rules. Using BIM, all the stakeholders can communicate and exchange information in real time using a virtual environment. This means that information can be shared without the need to use written documents as all of the actors have direct access to a centralized model. This ensures that all stakeholders have immediate access to the latest version of any document electronically. The design information that used to be conveyed and detailed in written form now appears directly in the model. This has the added advantage of being easily updated during the project's life cycle, depending on the needs of the stakeholders. BIM itself rapidly developed beyond 3D CAD into radically different paradigm of modelling buildings and a process of sharing information and making decisions in a more collaborative and integrated way. BIM is now giving way to a broader digitalization of the design, construction, and operation of the built environment.

The concept "Building Information Modelling" appeared in January 1975, in a paper by Eastman called "The Use of Computers Instead of Drawings in Building Design" (Eastman, 1975). Since then, it has gradually become a fundamental tool which is able to respond to the growing need to optimize processes, procedures, and decision-making. In 1994, Latham wrote about a new development known as "Knowledge Based Engineering" (KBE), calling it the "technology of the 21st century". He describes the potential to view advanced CAD models and see "all aspects of the design, manufacture, assembly and use of the product . . . presented in one entity" (Latham, 1994). This description is a clear precursor to what is now known as BIM. Dainty et al. suggest that it was not until the early 2000s that the acronym BIM first appeared in the industry (Dainty et al., 2017).

Despite ArchiCAD already being well established, it was not until the last 1990s and early 2000s that BIM began significantly to take off. This was when Nemetschetck's Open BIM (1997), Bentley's Triform BIM (1998), and Revit (2000) were all released as readily available software packages for the development of BIM (Smith, 2014).

The B for building is being dropped, with the recognition that the concepts behind BIM are also applicable to linear infrastructure and other types of built environment. Even the central concept of *information management* is giving way to a plethora of digital technologies which all work together to support the design, construction, and operation of the built environment.

1.5.2 BIM particularities

Construction drawings printed on paper are still predominantly used on construction sites. In design offices, the majority of CAD has been 2D – based on the two-dimensional cartesian coordinate system (Daniotti et al., 2020). Advances in computing have made modelling and drawing in 3D more feasible than ever before. It is now possible to add colours and textures to 3D models, to produce photorealistic renderings. 3D BIM also enables integrated parameterized information.

Time fundamentally constitutes a fourth dimension alongside the three space dimensions of our physical world. 3D models can be transformed into 4D models by linking 3D objects to activities in the construction programme. In this way, BIM can represent the sequencing and scheduling of construction, allowing clear communication of the construction process to stakeholders, better planning and avoiding of space-time clashes, and easier monitoring of progress against plans (Aouad et al., 2006).

Building on the four dimensions of the perceivable universe, Aouad et al. (2006, p. 5) proposed a concept called nD modelling, allowing for an "nth number of design dimensions" in a model – an undefined number of dimensions. This concept of having different dimensions, although there is not a global consensus among experts, is still frequently used. Using labels for distinct levels of detail is beneficial but using the term "dimension" in a metaphorical sense in a context where it is also used literally should be done with caution (Koutamanis, 2020). Dimension is only used in a literal sense up to 4D; beyond that, it is a likely metaphor for parameters or attributes linked to 3D objects in a model. The number of dimensions that the literature defines varies, with some sources going only up to 4D but others going up to 10D BIM (Software, 2018). The fifth dimension is commonly accepted as cost. 5D becomes possible to quantify and estimate costs for each task, creating a more precise budget and better cost analysis over time. Beyond 5D, however, there is little consensus over what each dimension is. These dimensions can include energy performance/sustainability (6D), as-built information, operation and facility management/life cycle maintenance (7D), quality, health and safety (8D), lean construction, and construction industrialization (Perera et al., 2017).

As an automated tool, BIM uses dedicated software with specific file types. One of the main concepts of BIM is the standard/specification called Industry Foundation Classes (IFC) which specifies the file type that is created. IFC was idealized by the International Alliance for Interoperability (IAI) and was followed up by the IAI successor, BuildingSmart International. BuildingSmart International was created to standardize BIM software files. IFC standardizes electronically, in an open format, all the parameterized and graphic information of objects in a single file format. It is a non-proprietary data model that creates and promotes interoperability between different modelling software and makes it possible for all the stakeholders to work and share data in the same model irrespective of the software application they use (Kamardeen, 2010).

Despite its obvious advantage, BIM is not yet as widely present in construction although its presence is growing fast. The complexity of BIM and the intricacies of its implementation gave rise to the need to categorize different levels of

BIM maturity which map out a pathway for BIM implementation. This was useful, for example, when the UK Government mandated BIM. The most used categorization in the United Kingdom is the Bew-Richards BIM maturity model which breaks down BIM maturity into four levels, sometimes known as the "slope" or the "wedge". Level 0 refers to paper-based CAD and is arguably not BIM at all. Level 1 refers to 3D models created in isolation and shared on a common data environment (CDE). Level 2 adds the element of close integration, through discipline-specific information-rich models that are created in a federated architecture but are interoperable and enable integration. Level 3 is known as fully integrated BIM whereby one single collaborative model can be edited by all disciplines at once as it is updated in real time. In fact, BIM is currently implemented to varying degrees in different works/places, ranging from level 0 to level 3 of implementation.

1.5.3 Level of development

BIM is object-based, and there are two alternatives for the creation of objects: either a new object is modelled and created "from scratch", or an object is retrieved from software industry libraries, for example, the NBS National BIM Library or BIMObject. When an object is modelled it can have different levels of detail and development. The Level of Detail relates to the detail of the visual representation in the model element, while the Level of Development (LOD) relates to the quality of information the object gives. More recently the term "LOIN" is used to refer to "Level of Information Needed".

Let us focus on the LOD, which varies a lot, depending on the phase of the project. The construction phase requires a less detailed LOD, which gradually becomes more detailed throughout the project, until it becomes a very detailed LOD in the maintenance phase (Cassano and Trani, 2017). The LOD levels proposed by the American Institute of Architects are:

- LOD 100 – the object is represented in the model with a generic representation such as a symbol and has parameters such as area, volume, and location;
- LOD 200 – the object is developed with more detailed information about dimensions and location;
- LOD 300 – the object is represented with precise information about dimensions, location;
- LOD 350 – the parameterized and more detailed object is enabled with an interface with adjacent objects;
- LOD 400 – the parameterized and interfaced object is provided with details about manufacturing, assembly, installation, and maintenance;
- LOD 500 – the object is represented as-built, with the necessary information about O&M.

The LOD 100, 200, and 300 are usually used in the design phase, the LOD 400 for the construction phase, and the LOD 500 for the O&M phase.

1.5.4   Potential advantages of BIM

BIM has emerged as the cornerstone of the digital transformation of construction, facilitating a more collaborative approach to project management, design, and execution. BIM technology has proven its worth by enhancing communication, cutting costs, reducing rework, and diminishing resource waste. Research suggests that BIM is a vital asset for civil construction in all its phases (from pre-construction to O&M and end of life; Clevenger et al., 2014). Efforts to use and transfer information via BIM for the installation phase are still (Wetzel and Thabet, 2015) embryonic with much less documentation regarding BIM in the O&M phase than in other phases (Cao et al., 2017). Perhaps because of this the focus of research seems to have shifted from the early stages of the construction life cycle to the O&M phases. BIM tools have been gaining increasing importance as an integrated management tool in the elaboration of Architectural, Structural, and Mechanical, Electrical and Plumbing (MEP) work. There are a number of advantages including:

- BIM provides a digitized 3D (Figure 1.1) and parameterized view of the design minimizing the need for large drawings;
- BIM largely anchors non-geometric information to the 3D model, and so is visual and removes language barriers which improves communication between all the players (Choe and Leite, 2017);
- BIM is easy to read, making it more accessible for those less able to read technical drawings or text;
- BIM entails the creation of a virtual environment where issues can be addressed in a common automated environment (Figure 1.2), thus optimizing and increasing the reliability of communication and information sharing, while minimizing information loss;

Figure 1.1 3D view (PERI Group, 2024)

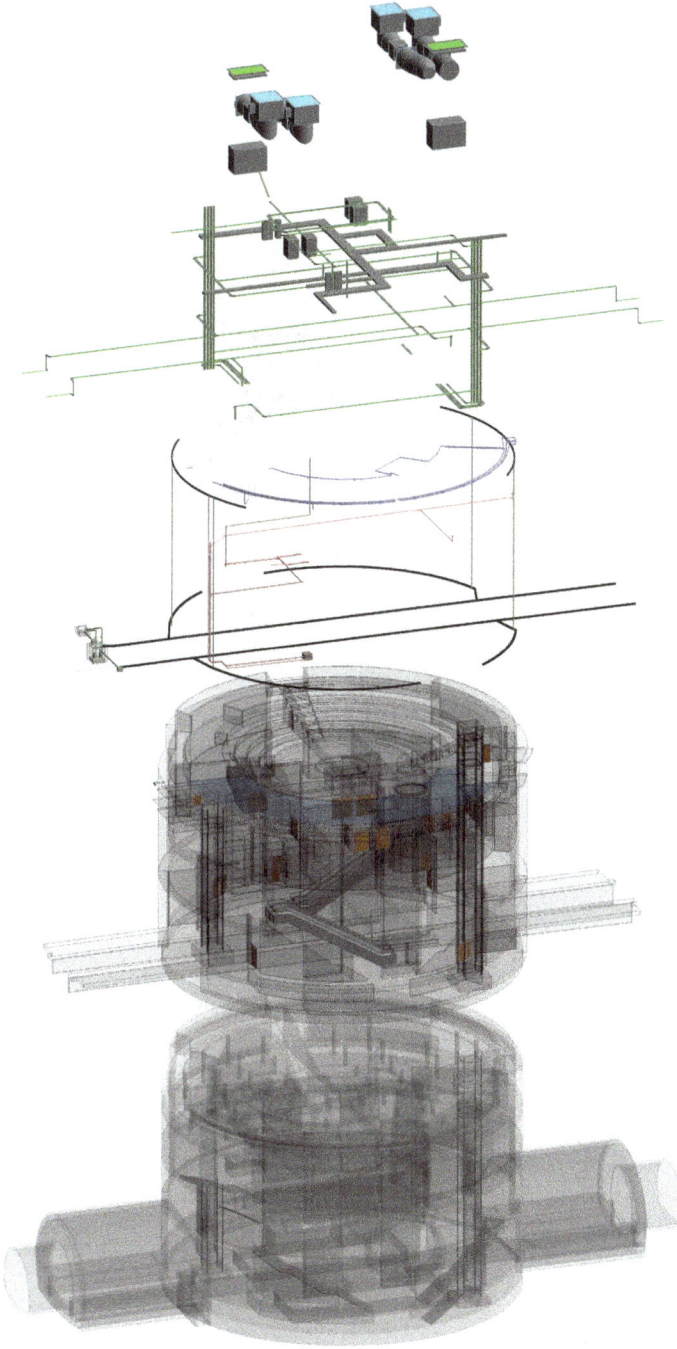

Figure 1.2 Common environment (Tender, 2024)

- BIM enables a reliable and validated information flow;
- BIM incorporates technical characteristics and specifications in each of the modelled elements;
- BIM speeds up decision-making, since it requires less time to obtain detailed results;
- BIM enables the visualization of interaction between different specialties and elements, since complex views and details are easily made;
- BIM decreases the number of human errors in graphic modelling;
- BIM minimizes work conflicts and incompatibilities before they even start (Alomari et al., 2017);
- BIM overcomes humans' limitations in terms of forecasting and interpreting different possible scenarios;
- BIM reduces manual amendments to the design;
- BIM enables immediate clarification of areas of uncertainty on-site avoiding the need for many paper documents;
- BIM allows comparisons between what is planned and what is accomplished.

Modelling can be upgraded using tools based in new surveying techniques such as laser scanning techniques (performed by ground equipment or drones), LiDAR, RGB-D sensors, and photogrammetry. These create a so-called point-cloud as a way of setting data points with 3D coordinates that precisely describe the geometry of real-world objects. 3D point-cloud data can provide accurate geometric data with a high data acquisition speed (Wang, 2019) which makes it very useful in construction settings.

BIM has not yet reached its full potential and is one of the drivers of the digitalization of the construction industry. BIM is arguably the most developed and used digital technology in the construction sector, and there is a relatively high average adoption rate of BIM.

The current business paradigm implies that a company must be competitive and dynamic, using strategies that allow it to make cost-effective decisions in a timely way. Applied to the construction industry, this means that the sector can and should adopt tried and tested tools such as BIM to support decision-making. The integration of BIM in this sector is increasing steadily, and stakeholders will need to adjust to the growing number of regulations and requirements on the use of BIM worldwide.

A lot of the focus of BIM is on the public sector and large-scale projects, but that does not mean that best practice is not transferable to smaller schemes (Mordue and Finch, 2019). Simple projects may be modelled just as complex projects. However, simple projects are often just as complex and demanding as a "large project", as complexity does not just relate to the size of the asset; they still require coordination and management of information, but perhaps to a different level of detail (Mordue and Finch, 2019).

1.6 BIM for OSH

Over the years, digital technologies have played a pivotal role in transforming the landscape of OSH. Overall, these technologies have revolutionized safety measures and enabled proactive safety planning, real-time monitoring, and risk reduction associated with various construction processes, ultimately enhancing the safety and well-being of workers and improving project results. These technologies still have low uptake in most AECO organizations, although there are already some cases where new technologies have been successfully implemented (Health and Safety Executive, 2018).

The General Principles of Prevention established in Directive 89/391/EEC (European Commission, 1989) stress the urgent need to take into account the latest technologies and the state of the art, namely in terms of Information Technologies. Requirements of Directive 92/57/EEC also refer the need of approaching new technologies for developing OSH. It can be said, based in these Directives, that the emerging technologies for OSH in construction industry, are more than a legal obligation as the project supervisor, or where appropriate the PO, shall take account, measures such as adapting to technical progress and developing a coherent overall prevention policy which covers technology, during the various stages of designing and preparing the project, in particular when architectural, technical, and/or organizational aspects are being decided.

The literature shows that the lack of OSH information in digital form is one of the many factors at the origin of the low performance of OSH management in the construction industry.

Over the years there have been many attempts to automate the process, with some being more successful than others. The use of BIM in construction is now widespread, especially in architecture, structures, or MEP. It is not, however, so widely used for OSH purposes. Although new technologies have been mostly used for as-built models, they have, as indicated by early studies of related aspects, a potential use in the OSH field (although they have yet to be comprehensively and coherently considered with respect to the specific challenges of OSH) and their use has been linked to an improvement in safety conditions and to a decrease in the accident rate in recent years (Martínez-Aires et al., 2018).

In this new digital age that increases global competition the importance of having OSH information included in the global digital data set of the project during its life cycle will increase and new management techniques are to be expected (Sulankivi et al., 2010). It is to be expected that, as people realize the importance of OSH in the digital data environment and increasingly recognize the advantages of using BIM, the number of related studies will also increase. Large companies in countries like the United Kingdom, the United States, and China have already begun this journey.

The inclusion of OSH information in digital data through BIM can assist the automation of occupational risks prevention in the life cycle of the project by creating optimized processes of collaboration, communication, and coordination between all stakeholders. It can be used to set properly tested standards and good practices within a collaborative environment using parametric modelling which allows active and efficient sharing of information between all stakeholders in the project. All of this will bring about a more efficient, profitable, and sustainable construction process.

The use of BIM to include OSH information in digital data has increasingly interested researchers in the field of construction (Zou et al., 2017). Research into BIM for OSH has been increasing (a trend that is clear from the number of published research articles and their citations since 2017). Aguilera analysed articles published in 11 countries on the use of BIM methodologies applied to OSH in construction, noting that 89% of articles were published between 2012 and 2016, with special attention from 2013 onwards (Aguilera, 2017). The country with the highest percentage of published articles is the United States (39%), followed by China (20%) and South Korea (15%).

Several authors have published reviews about the theme (Table 1.1).

Several influent authors have published papers about the theme: Billy Hare, John Gambatese, Jochen Teizer, Matthew Hallowell, Roger Haslam, Alistair Gibb,

Table 1.1 Published literature reviews

Year	Author	Name of paper
2011	Azhar	Building Information Modelling (BIM): Trends, benefits, risks and challenges for the AEC Industry
2012	Zhou et al.	Construction safety and digital design: A review
2015	Ganah et al.	Integrating Building Information Modelling and Health and Safety for Onsite Construction
2016	Hallowell	Information technology and safety: Integrating empirical safety risk data with building information advantage, sensing, and visualization technologies
2017	Zou et al.	A review of risk management through BIM and BIM-related technologies
2018	Martínez-Aires et al.	Building information advantage and safety management: A systematic review
2019	Akram et al.	Exploring the role of building information advantage in construction safety through science mapping
2019	Jin et al.	A science mapping approach-based review of construction safety research
2019	Aguilar et al	Review of health and safety management based on BIM methodology
2019	Mihic et al.	Review of previous applications of innovative information technologies in construction health and safety
2020	Fargnoli et al.	Building Information Modelling (BIM) to enhance occupational safety in construction activities: research trends emerging from one decade of studies
2021	Tender et al.	Improving Health, Safety and Well Being data integration using Building Information Modelling – a literature review
2022	Hoeft and Trask	Safety Built Right in: Exploring the Occupational Health and Safety Potential of BIM-Based Platforms throughout the Building Lifecycle

Kristiina Sulankivi, Maria Dolores Martinez Aires, Salman Azhar, Yang Zhou, and Arto Kiviniemi. Most of the articles that have been published are focused on planning and design phase, specifically on hazard recognition and prevention. Those various studies carried out in different countries have identified several advantages in the use of BIM in the OSH area. Although they have shown BIM to be effective for OSH, the vast majority of research into BIM for OSH is still very conceptual and based on simplified hypothetical buildings ("square shape building") with a low level of complexity and often only tested in laboratory conditions (Mordue and Finch, 2019). In addition, most of the research was related to site planning and fall from height risks, and did not cover any other type of risks such as roll over and crushing. Thus, the results are interesting but do correspond to practical solutions that can be effective and easily implemented (Mordue and Finch, 2019). Finally, a word of caution: it is important to bear in mind that the usefulness of BIM for OSH is heavily dependent on the quality of the BIM model itself. If the model's information is incorrect, then all of the resulting OSH outputs will become incorrect as well.

Some authors, such as Kamardeen, argue in favour of addressing the use of BIM for OSH based on 8D BIM (Kamardeen, 2010). However, it could just as well be integrated into 7D BIM, that is, sustainability in the workplace. Our opinion is in favour of having a separate dimension for OSH, as supported by Kamardeen, as organizations cannot strive to be sustainable if they do not take into account the OSH of their employees.

Standardization is essential for the implementation of new technologies in AECO, and it is expected to be one of the ways for BIM to be widely used and uniformly accepted (Bragança et al., 2020). The need for the standardization of BIM also arises from the accelerated growth of its use coupled with the need for construction projects to standardize and maintain acceptable quality levels while reducing the risk of information loss.

National authorities worldwide, especially the most mature in digital implementation, have been realizing the importance of integrating new technologies in OSH management, in particular that BIM can help in identifying and mitigating risks in an early stage (Zou et al., 2017). It should be noted that the lack of integration of methodologies can create coordination problems and pave the way for the increase in occupational accidents (Cortés-Pérez et al., 2020). Despite several efforts, the standardization of BIM for the prevention of occupational risks is still in its infancy in most countries. Countries like Finland, the United Kingdom, and the United States already have some institutional approaches to using and sharing OSH information through BIM. The efforts made are of great relevance, and date back to 2012, with the launch of the "Common BIM Requirements" in Finland. This document refers to safety, namely in terms of the need for risk analysis, the gradual introduction of safety information, site modelling with specific rules and predefined objects, and the introduction of OSH information for the O&M phase. It was also in Finland that R&D of this area was born, namely with the VTT Institute R&D Projects TurvaBIM "Building Information Model (BIM) promoting safety in the construction site process" (2007–2009) and BIMSafety "BIM-based

Safety Management and Communication for Building Construction" (2009–2011) financed by Tekes – the Finnish Funding Agency for Technology and Innovation and The Finnish Work Environment Fund and some contractors as Skanska. This research was made in a partnership between VTT, the Tampere University of Technology, and the Finnish Institute of Occupational Health.

The United States were also pioneers, with the "Building Information Modelling Site Safety Submission Guidelines and Standards" (New York County Buildings, 2013) being created by the New York City Department of Buildings. This document describes the requirements and procedures for preparing, submitting, amending, and reviewing electronic plans for the site with subsequent submission and approval by the municipality with the provision of a specific object library for the site. Next in this journey for standardization comes the UK standard "PAS 1192–6 – Specification for collaborative sharing and use of structured Health and Safety information using BIM", published in 2018.

This really is a milestone in this area, since it is the first regulatory document worldwide to address the use and transmission of preventive information using BIM, and it still remains the most advanced document in this field (Bragança et al., 2020). It details the roles and duties of each stakeholder in a project stressing the need to deal with prevention right from the design phase. This standard does not unequivocally define the method to be used for incorporating health and safety information into BIM (Cortés-Pérez et al., 2020); instead, it provides three alternatives to accomplish this (better explained further). Additionally, to these national-level regulatory documents in place, ISO 19650–6, dedicated to OSH, was launched in early 2025. Despite these good examples the fact remains that, in most countries, using BIM is neither legally mandatory nor regulated which results in a lower general maturity level than in the countries mentioned earlier. By understanding the challenges that the countries that lead these processes had to overcome, and using them as references and benchmarks, it will be possible to minimize the time needed for the correct implementation of BIM.

Categorization of OSH areas to apply to digital data has been explored by several authors distinguishing the areas in which BIM can be useful in different ways. In this book, the authors choose to divide it into the following areas that are described in Council Directive 92/57/EEC on the implementation of minimum safety and health requirements in temporary construction sites: documental/contractual management, risk identification and assessment, training, site planning, tasks planning, monitoring, emergency planning, and work accidents investigation.

A number of advantages of using BIM for OSH have been identified by previous research, with most of them being aligned to European regulations. These include:

- BIM appears to be a valid instrument for integrating OSH information in planning (Tender et al., 2017b), and its increasing implementation is changing the way safety can be addressed (Martínez-Aires et al., 2018);
- BIM is the best tool to improve safety performance in the construction industry (Alizadehsalehi et al., 2017);

- BIM is already seen as a "catalyst" for improving safety concerns by taking into account the dynamic work environment and working conditions (Ganah and Goudfard, 2015);
- the implementation of BIM, from the point of view of safety management, saves time and effort (Martínez-Aires et al., 2018);
- BIM can effectively manage the links between design and safety on-site (Kamardeen, 2010);
- the use of BIM in the design phase aimed at providing good life cycle management allows to considerably reduce the costs of the construction life cycle, especially because the O&M lasts much longer than the design and construction phases (Matarneh et al., 2019);
- the automation provided by BIM will help to mitigate the problem of OSH professionals not having the time or the resources to conduct detailed analyses of each construction site (Mihić et al., 2019).

It is already discovered that the lack of BIM integration methodologies can create coordination problems and pave the way to increase the number of accidents at work (Cortés-Pérez et al., 2020).

There are several factors that support the implementation of BIM-based actions in practice:

- the actual high level of software interoperability, which allows the application of the BIM model for various purposes, such as visualization or simulation, regardless of whether the open-source, standardized format-supported OpenBIM concept or the ClosedBIM concept using a unified software environment is chosen;
- the widespread use of cloud technologies and web-based environments, coupled with accessibility on mobile devices, enables real-time access to the safety-related information for participants;
- the high level of usability and user-friendliness aids in the easier and intuitive adoption of digital solutions.

Several authors say that BIM for OSH should be developed, stressing the need for further research. Although current approaches using BIM appear to be focused on specific areas, such as risks of falling from a height (Guo et al., 2017), several authors mention the need for research to widen its scope of application such as:

- "there is a need to combine the approach via BIM and the traditional to improve practical applicability" (Zou et al., 2017);
- "lack of advertisement for BIM by the construction professionals and industry" as a factor influencing BIM unawareness by clients (Olugboyega and Olugbenga, 2018);
- "there is no comprehensive view of recent research on risk management based on BIM as a comprehensive whole and no study focusing on the relationship between digital technologies and traditional methods for managing risk [3]";

- "the future agenda, creating and reusing consistent digital information by stakeholders throughout the life cycle" (Arayici et al., 2012);
- "must integrate health and safety information with BIM information to provide a rich database for O&M systems" (Matarneh et al., 2019);
- "[t]here is an urgent need to combine BIM methodologies with traditional safety management processes (KarimiAzari et al., 2011)";
- "create a mechanism in FM for application of H&S issues 'and' more real-world case studies, focusing on assessing the current challenges that hinder the implementation of BIM in FM" (Matarneh et al., 2019);
- "more lessons learnt from practical cases to address the gap between the theory and realised benefits by industry" (Pidgeon and Dawood, 2021);
- "future studies should focus on continue exploring new technologies for OSH implementation in real practical cases" (Tender et al., 2022d);
- "dynamics and trends of BIM for OSH at European level and different National levels should be explored" (Tender et al., 2022c);
- further work should be considered which explore in more detail some of the areas identified and establish the benefits of using digital technologies to improve OSH outcomes (Tender et al., 2022a);
- is needed to explore the advantages that were identified in order to validate their theorical validity using suitable case studies (Tender et al., 2022b);
- there should be a focus on detecting weaknesses and threats, in order to try and find solutions for them (Tender et al., 2021);
- "incomplete technology transition from construction safety research into practice (Mordue, 2015)";
- "there is a gap between the theory and realized benefits at the application stage by industry (Pidgeon and Dawood, 2021)";
- "the adoption of BIM by the construction industry remains limited, showing a gap between policy and practice which may be paradoxical as the industry is in theory the prime beneficiary of digitalisation (Observatory', E. C. S. 2019)";
- "there is no universal instrument to observe and monitor the progress of BIM for OSH implementation, trends and dynamics in each particular country or at European levels (Tender et al., 2023)";
- "there is no organized repository covering the capture, storage and dissemination of lessons learned. This can lead to a widening knowledge gap that can affect future projects as regular collection of lessons learnt, and their meaningful utilization in subsequent projects are critical elements of success" (Anbari et al., 2008).

1.7   How to include OSH information in the model

The actual great challenge is how to integrate OSH information in digital data. It should be noted that PAS 1192–6 indicates three ways to integrate BIM information: graphic marking, parametric information, or COBie.

1.7.1   Graphic marking

Graphic marking is one of the most basic and common ways of integrating OSH information into the model's digital data. It consists of introducing a visual identifier in the graphic component of the model (this symbol will always be in that model whatever the way the model is viewed – as a plan, section, or elevation) that provides a visual cue about a risk that is present in that place or within that object by making a simple click on the mark to have OSH information. Marking can be also achieved with a 3D symbol. This symbol can have expressly written the risk.

This allows user to have immediate access, with no need to look for papers, to additional information about each risk. This symbol can have information about the risk and can also be linked to OSH documents such as a Safety Sheet.

1.7.2   Parametric information

The introduction of parametric information in the object is another alternative given by PAS 1192:6, and it means having OSH information in the parameters of the object. Usually, objects are parametrized with information about properties such as mass, volume, and dimensions. The OSH information that also can be included in the object parameters (project parameters or shared parameters) can be risks, preventive measures, safety perimeters, user manuals, manufacturers information, and so on. This could be done with project or shared parameters (Figure 1.3).

This also means that, in the O&M phase, detailed OSH information from the Health and Safety File will be directly embedded in the model and will be easily identified from the as-built model (Health and Safety Executive, 2018).

1.7.3   COBie

Construction Operations Building information Exchange (COBie) is a non-proprietary data format for monitoring asset information in the operational phase. It categorizes information about all the asset's characteristics through easily accessible

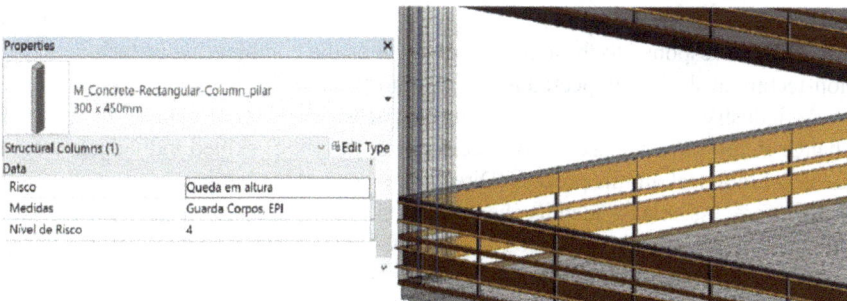

Figure 1.3 Safety parametrized information (Tender et al., 2017b)

spreadsheets. The information can be sorted by elements, spaces, or activities. Integrating OSH in digital data using the COBie is the other possibility offered in the PAS 1192:6 COBie tables. COBie contains an "Issues" item that reflects the IFC4 risk model recommended by BuildingSMART. The BuildingSMART IFC4 Property set "Pset Risk" recommends that the overall risk, likelihood, and consequences are scored on a defined scale. This item includes nine sub-items, including elements such as "Risk" (risk rating), "Chance" (risk assessment), and "Impact" (risk consequence) which can have links to other documents, for example, OSH procedures, instruction manuals, maintenance manuals.

1.8 Skills

The future of work is one of the most discussed topics worldwide due to exponential discoveries in the field of technology. Low productivity and skills shortages lead to high construction costs, delays in construction projects, and poor sustainability practices in the construction sector (Turner et al., 2020). This makes skills an important factor when studying the implementation of BIM. There is a broad consensus that the advancement of automation and digitalization will continue to transform the skills requirements for employees in the coming years (Deloitte, 2020) and that the need for competent workers is fundamental for the adoption and implementation of health and safety technology (Nnaji and Karakhan, 2020). The recent report "EU 2030 High-Tech Skills Vision for Industry 4.0" highlights new and innovative skills for construction professionals, who are in line with the paradigm of smart industrial specialization and digital transformation (Directorate-General for Internal Market, 2019). Although combining soft skills with technical skills is still a challenge that requires different approaches to training and life-long learning practices, requalification and updating to keep up with technology is an important part of maintaining valuable workforce of all ages (Directorate-General for Internal Market, 2019). This fact will require an additional effort on the part of all involved in order to achieve training requalification if actors really want to respond as quickly as possible to the changes in skills requirements triggered by digitalization (Deloitte, 2020). BIM inevitably implies a need to update the skills of OSH professionals since a qualified workforce is a crucial factor for the success of its adoption. In response to these market challenges the combination of technical and non-technical skills is expected to be in high demand.

As industry is reconfiguring working practices (Farrell et al., 2020), the pace of change in the skills needed is also accelerating. A lack of skilled individuals would create a bottleneck in this process (Directorate-General for Internal Market, 2019). It is, therefore, essential to identify the required skills and propose a roadmap to upskill the workforce (Delgado et al., 2020). The UK Government is already seeking to address this, with the programme Digital Built Britain (Level 3 BIM Strategic Plan), which refers to "training the public sector client in the use of BIM techniques" as a key measure. It also states that the long-term success of their strategy would be underpinned by "an effective education and change management programme to enable the industry to develop necessary skills" (Digital Built Britain, 2015).

It is important to say that the quality of information in BIM is only as good as the skill and knowledge of the people who produced it (Mordue and Finch, 2019). Currently, even in BIM-mature countries, few design professionals have developed their BIM skills, abilities, and working practices to the point where they are able to focus on aspects such as OSH (Shih et al., 2012).

New technologies have put new emphasis on the skills described as "business", "human", and "soft" skills that machines cannot master. It is, therefore, essential to identify the required skills (such as analytical and critical thinking, flexibility, evidence-based decision-making in the face of uncertainty and complexity, acting within multidisciplinary and intercultural context, problem-solving, creativity, and innovation) and propose a roadmap (involving academic and industry approach) to upskill the workforce.

This will require additional effort from employers and individuals in order to achieve upskilling if they are to be capable of responding as rapidly as possible to changes in skills requirements triggered by digitalization (Deloitte, 2020). Therefore, BIM inevitably implies a need for the skills of OSH actors to be upgraded to create "digital workers".

Table 1.2 collates some important key competences OSH professionals need to have to be effective in their tasks. The competences identified are supplementary to those mandatory by legislation.

Table 1.2 Competences

Theme	Competences
General skills	English as a foreign language
	Solving hardware/software basic problems and conflicts
Specific knowledge/skills	Understand copyrights and licenses
	Know the general/specific concepts and application of each new technologies
	Work with remote teams in specific tools (CDE) through Teams, Zoom, etc.
	Edit/create, through 3D/4D tools and plugins, site and activities planning
	Know the potentialities of simulators for training
	Security, identification of vulnerabilities
Physical skills	Body coordination and flexibility (arm-hand finger steadiness/dexterity)
	Reaction times
Socio-emotional	Feeling physically safe, secure, and healthy
	Openness to learn and interact with new technologies in workplaces
	Active learning, flexibility, adaptability to new technologies procedures
	Good interpersonal/intercultural/collaborative relationships
	Time management
	Innovative thinking and creativity
	Emotional intelligence – understanding someone's online emotions and moods
	Critical thinking and analysing online arguments
	Conflict management in dilemma Safety/Production
	Social intelligence – perceptiveness, persuasion, and negotiation

However, it is difficult to predict exactly how the importance of each specific individual competency will change, or which competencies have a low capacity for automation and will therefore be in demand in the market (Deloitte, 2020).

When analysing competences, it must be understood that, within people who work with BIM and other digital technologies, there are two totally different groups: the generation born in the end of the 1990s and beginning of the new Millennium, the so called "millennials", who have actually always been immersed in several kinds of digital technologies in almost every aspect of their lives, facilitating the acquisition of technical skills, and the so called baby-boomers generation, born between 1946 and 1964, who are often reluctant towards technology and may find, for example, video games unrealistic and a frivolous way of learning (Dickinson et al., 2011). These two totally different ways of seeing digital technologies have advantages and disadvantages.

Chapter 2

Document/Contract management

2.1 Traditional approach

Until recently, there was no central, common environment where all stakeholders could share project data as "a single version of the truth". The increasing complexity of projects requires the contractualization of procedures and collaboration between interested parties, as information interdependencies necessitate more formal setting out of communication and documentation requirements. Additionally, current OSH systems can prove inadequate as they are often translated into an excessive list of extensive, unstructured, and difficult-to-understand procedures (Tender et al., 2017a). They are often considered tedious by those who should implement them with little integration into the work planning. All of this limits the possibility of identifying and analysing dangers before the construction phase begins (Azhar and Behringer, 2013). This is exacerbated by the information related to the whole life cycle of a project being usually taken from several unconnected and uncategorized analogue and digital sources. This results in a very high probability of misplacing hard copies of documents. Also, stakeholders do not always know whether they are working on the most up-to-date version of a document. The sources used by OSH professionals (paper, images, recordings, etc.) come in several file formats (.xls, .doc, .dwg, .rvt, .ppt, etc.), contain hundreds of attributes, and often lack any structure. Therefore, it takes time to capture, analyse, integrate, and interpret, making it difficult to fully decode significant information. All of the above constraints can lead to duplications, overprocessing, rework, waste of time, or information inefficiencies. In addition, 2D drawings are usually hard to read and interpret, and the presence of foreign workers can create an additional language barrier which becomes an obstacle to OSH implementation. This can be further complicated when POs are not even aware of what information they are going to need for each phase of the life cycle of the asset. An added hurdle comes with remote construction projects, in which those involved are located in different location or countries (Arayici et al., 2012). This ineffective communication among project stakeholders can lead to a poor collaborative working environment, which

DOI: 10.1201/9781003615217-2

in turn can lead to confusion, misinterpretation, and ignorance with respect to critically important information, and result in severe impacts on project performance in terms of delivery quality, construction schedule, project cost, and site safety (Zhang et al., 2020). The life cycle of construction projects is treated "in chunks" by current construction contract models (design, construction, and O&M phases are considered separate phases with reduced integration between phases) which makes information hard to share effectively among stakeholders. Therefore, the construction industry needs a robust solution to facilitate the coordination between stakeholders and enhance communication across disciplinary knowledge boundaries (Zhang et al., 2020).

2.2   BIM approach

Cloud-based solutions or collaborative environments, known as a CDE, consist of platforms (e.g. an extranet, a project-dedicated server) for automated, structured, centralized, and open document management. These provide a standard-compliant environment to specify, collect, safe-keep, store, present, and manage BIM information in a spirit of collaboration (UK Government BIM Working Group – CDE Sub Group, 2018) during the project's life cycle. This allows information to be used and shared by all involved in the process by gathering traditionally non-centralized information in a single place. This way, all the members of the project can collaborate in a single platform working in the same model.

CDE information is usually split into four areas (Caires, 2013): Work in Progress, Shared Data, Publication and Document Issue, and Archive, which assists in organizing documents. CDE allows users to identify, mitigate, manage, and communicate risks throughout the project (Health and Safety Executive, 2018) by improving the processing/transmission flows of information as each information extract goes through several approval processes. In order for the CDE to work, the files must not need proprietary software or hardware to be read, and this is one of the main principles of IFC working.

Regarding OSH data, the CDE has the following advantages:

- it provides a robust, structured, and managed system to record and register health and safety information throughout the project life cycle (Health and Safety Executive, 2018);
- it facilitates communication between stakeholders;
- it provides users with the early opportunity to identify hazards, mitigate risks, and share health and safety information, increasing the efficiency of project delivery (Health and Safety Executive, 2018);
- it acknowledges design risk management as an integral part of the design process and allows the principles of risk prevention to be implemented in the development of the design solution (Health and Safety Executive, 2018);

- it enables a more peaceful connection between design, construction, and O&M phase;
- it facilitates pertinent archiving and document management including Health and Safety Plan in the project phase; Health and Safety Plan development in the construction phase; specific safety procedures; approval of subcontractors, workers, and equipment; Health and Safety File.

The contracting phase is arguably the first phase where BIM becomes relevant for establishing how OSH information can be inputted in digital model data. BIM can be used to organize the usually large amount of information that may have to be made available in the tender phases. It makes perfect sense for the model and the information to be used in an appropriate way throughout the life cycle of the project. This is accomplished with the following:

- The Employer's Information Requirements (EIR), which is a new document stemming from the BIM approach, expressing at the tendering stage the PO's ambitions in terms of using BIM. The document sets the minimum requirements to be met regarding the use of BIM by subcontractors (designer, contractor, inspection, etc.). The EIR's content is defined by PAS 1192:2 and ISO 19650, and it is intended to be straightforward, clear, fair, impartial, and objective. After the contracting phase the EIR then serves to assess the fulfilment of technical (software platforms, exchange format data, coordinates, type and level of information and detail, training), management (standards, roles and responsibilities, data planning, security, clash-detection process, collaboration processes, health and safety management, systems performance, compliance plan, building management strategy), and commercial (data and deliverables, skills assessment) requirements.
- The BIM Execution Plan (BEP) is another new document introduced by BIM, and it is the contractor's response to the EIR defined in the tender phase to explain how the information modelling aspects of a project will be carried out. The BEP has two different phases: pre-contract BEP and post-contract BEP. In the BEP, the contractor states the methodology used to implement each of the requirements stipulated by the PO and commits to the strict compliance with the EIR stipulated in the tender phase.

When it comes to OSH information, both the EIR and the PEB must ensure that all preventive information to be included in the OSH documents is properly transposed to the model. This cross-checking of information between the EIR/BEP and the OSH documents is paramount for the correct understanding of the PO's intentions regarding OSH for the life cycle of the building.

The OSH documents that will be created must comply with legal, normative, and contractual requirements, but caution must be taken in order to avoid creating an excessive, unnecessary, and unreadable amount of paperwork.

The use of BIM for OSH cost management enables quantification and budgeting through simulation models. The advantages of this include:

- analysis of costs during all phases of the life cycle which will be reflected in budget estimates, financial control, and costs;
- early and automatic measurement of OSH equipment (construction fences, signage, guardrails, etc.) which enables a speedy approach to cost estimates for each activity (Martínez-Aires et al., 2018), decreased time spent on budgeting, and also a reduction in future disputes regarding financial terms;
- easier correction of costs and budgets whenever a change is made in the project in terms of construction solutions, materials, equipment, and labour, among others, also allowing to compare, in terms of costs, different scenarios and solutions (Eastman et al., 2008);
- automated extraction of quantities (something impossible in the traditional process) which allows for a fast listing of preventive elements installed or assembled at any given moment.

This is particularly important when there are changes in the construction site structure, equipment, or work planning. If unit costs are then linked to that list, a complete budget can be swiftly assembled (Tender et al., 2018b). This reduces the time spent in quantity take-off and estimation from weeks to minutes and increases the accuracy of results, which reduces the possibility of future financial disputes.

- reduction in the time spent in quantity take-off and estimation from weeks to minutes (Kamardeen, 2010);
- at any time and at any stage of the work know what material is needed and forecast the estimated final cost related to each activity which optimizes financial management (particularly in relation to suppliers; Tender et al., 2018b);
- easier correction of costs and budgets whenever a change is made in the project in terms of construction solutions, materials, equipment, and labour. This allows the comparison of the costs for different scenarios and solutions (Eastman et al., 2008);
- automation of cross-referencing protective equipment to its cost will allow separate costing of items against the specification instead of giving the usual "bulk price" approach. This solves the existing problem of lack of quantification of items related to prevention which has the potential to minimize unfair competition (Tender et al., 2018b);
- virtual testing of various solutions and scenarios from an OSH perspective allowing the projection of financial scenarios to optimize and control costs;

- mechanisms to understand where significant costs occur and also when they occur in the timeline;
- improves the accuracy of estimates (Kamardeen, 2010);
- reduction in costs across the construction life cycle including the O&M phase which lasts much longer than the design and construction phases (Matarneh et al., 2019).

Chapter 3

Risk identification and assessment

3.1 Traditional approach

Construction workers are constantly exposed to several risks in all the phases of the construction life cycle, for example, falls from height, electrical shock, crushing, cuts, and bruises. The integration in the design phase of mitigation measures addressing these risks depends on the PO. It can bring gains in technical, logistic, and financial terms. Several studies have demonstrated that poor or even non-existent planning in the design phase played a significant role in more than half of the occupational accidents in construction (Martínez-Aires et al., 2018). European Directives require that a project's risk analysis begins at design and indicates that risks must be identified as fully as possible during the design and planning phases (Benjaoran and Bhokha, 2010). Kamardeen stated that the ideal time to influence construction safety is during the inception, concept design, and detailed design phases (Kamardeen, 2010). In these phases, designers can influence construction safety by making better choices in the design stage of a project (Marefat et al., 2019). In the inception and concept phases, designers can influence construction safety by making safer choices in terms of geometry, material, or techniques to be used. As the project progresses, it becomes harder to influence safety. European Directives also stipulate that the designer shall inform the contractor about any residual risks that could not be eliminated during the design phase so that the contractor knows that they will have to plan for appropriate measures to reduce the likelihood or impact of those risks. It is up to the designer to take a preventive approach in the evaluation and elimination of risks that may occur during construction. In order to do so the designer must choose design solutions having safety as a criterion in addition to the cost or performance (Tender et al., 2015). The identification of risks and their mitigation is subject to the designer's ability to identify and predict potential risky situations. The more detailed the design and the more consideration the design has for safety issues (Hossain et al., 2018), the simpler it will be to understand the potential conflicts regarding OSH (Alomari et al., 2017).

In most countries, OSH is usually not adequately included in the design phase with a few exceptions. OSH is often only addressed after the design is already concluded. Most of the time, the designer does not even have the knowledge or

DOI: 10.1201/9781003615217-3

experience to address construction and O&M risks in the design phase. This means they are unable to identify the impact that their design solutions will have in the construction phase (Zhang et al., 2012) and will not be able to act on these issues. As a result, the opportunity to reduce the construction risks in the design phase will be lost. This places the actors in the construction phase in the very delicate position of trying to minimize risks that should have been handled previously. Often, risks that could have been prevented by adopting other design options are hard to handle on site, and it may even be impossible to lower them at that stage (Tender et al., 2015). On the one hand, is that designers are often unaware of their preventive responsibilities and assume that OSH has nothing to do with them because it is the sole responsibility of the builder. This means that they seldom seek any feedback on the impact of design work on safety risks during construction and operation. On the other hand, there is a bulk of knowledge accumulated on job sites through years of experience that is passed tacitly from generation to generation of on-site work-ers but that cannot be fully shared with designers (Yuan et al., 2019). Literature shows that there is a lack of appropriate tools and resources to assist designers with addressing construction safety (Ku and Mills, 2010). It appears that safety in the design phase is less mature than that in the construction phase (Mordue and Finch, 2019). It should also be noted that, in most cases, there is no analysis of the existing risks for each project; instead, a generic base model is often created which is then used for all projects (Tender et al., 2018a).

After risk identification, the next phase of the risk assessment process is risk estimation. This assessment of risks is often very difficult especially in potentially dangerous activities regarding both the assessment of the probability of occurrence and the estimation of its consequences. It is also remarkably subjective because of the subjectivity linked to the upstream estimation of variables, usually because of a lack of reliable data and also because the documentation that does exist is fragmented and seldom accompanied by robust analysis.

The next step is assessing the risk by comparing it with benchmarks to establish the degree of acceptance and tolerability. This allows for the provision of accurate information for the employer to take the appropriate preventive measures through safety management to minimize/eliminate risks. Establishing the benchmarks in terms of the risk acceptance and tolerability criteria is one of the key factors to guarantee the reliability of the risk assessment process (Ale et al., 2008). It should be noted that as technology evolves the threshold of acceptability of the level of risk by the general population decreases [19]. Situations that were acceptable sev-eral years ago are now considered unacceptable, so approaches to OSH risk tend to be more rigid and conservative.

In terms of establishing preventing measures, the workers might not know or be aware of when these safety measures would be needed (Benjaoran and Bhokha, 2010).

Additionally, currently, detection and checking for issues that can give rise to risks is traditionally done manually. This is time-consuming laborious work (Park and Kim, 2015) and, as it is manual, can also be error-prone. These rules are sometimes conflicting and incomplete, and the corresponding implementation is often limited by people's understanding, interpretation, and reasoning capability (Zou et al., 2017). The problem is that it is also extremely important activity. If incompatibilities are not detected and treated on time, they can have a significant impact on costs and deadlines. Avoiding errors in this process is paramount, which makes automation desirable (Park and Kim, 2015).

3.2 BIM approach

Several authors have researched the use of BIM for risk identification and assessment: Kamardeen focused on the design phase and on understanding the safety risks and consequences using automated risk detection (Kamardeen, 2010). Zhang developed an automated risk identification for the design and construction planning stages (Zhang et al., 2012). Using a case study, Arayci studied how BIM adoption helps to minimize communication problems (Arayici et al., 2012). Zhang put forward an automatic ontology-based semantic modelling of construction safety knowledge for job hazard analysis (Zhang et al., 2014). Choi explored workspaces to minimize congestion hazards (Choi et al., 2014). Hayne explored BIM potential for disseminating design for safety knowledge (Hayne et al., 2014). Wetzel focused on O&M phase BIM-based risk management (Wetzel and Thabet, 2015). Zhang developed a BIM-based Risk Identification Expert System (B-RIES) for tunnel construction (Zhang et al., 2016). Xiaer researched using BIM in the design phase for safety management improvement (Xiahou et al., 2016). Malekitabar studied the analysis of causes linked with safety rules based on BIM (Malekitabar et al., 2016). Hayne focused on the design phase, providing a mixed media approach combining video, audio, and animations stemming from the BIM model (Hayne et al., 2016). Shen and Marks worked on manual introduction of the risks associated with an element by assigning them a colour code based on their severity (Marks and Shen, 2016). Wetzel proposed a framework for safety information delivering to O&M workers (Wetzel and Thabet, 2015). Li focused on BIM-based risk identification in underground construction at the pre-construction stage (Li et al., 2018). Mihić studied the requirements for using risk and hazard databases based on BIM (Mihić et al., 2018). Deng explored risk assessment databases and used safety management modules based on the development of Revit platforms for identifying relevant risk sources (Deng et al., 2019). Perez focused on studying integration of the H&S risk assessment in building projects with BIM by considering a specific H&S subdiscipline in the model (Pérez et al., 2017).

For the design phase, one strand of research focuses on use of BIM as a data repository and finds that the most common interface of BIM with other new technologies was with various knowledge-bases and the use of ontologies, mostly for hazard identification through rule checking. Focusing on rule-checkers, software has the potential to encode rules and criteria by interpretation of legal, normative, regulatory, contractual requirements, limit values (qualitatively and quantitatively), and guiding documents (Mordue and Finch, 2019) and thus building models could be checked against these machine-readable rules automatically with results, for example, "pass", "fail", "warning", "unknown" (Zou et al., 2017). This search for non-compliances, or nonconformities, can be referred to as *rule-checking*. The rule-checking process is composed of four phases: rule interpretation; BIM model preparation; rule execution; and rule-checking reporting. If the check is "ok", which means that a certain set of criteria have been met, the model is allowed to be validated and guaranteed as suitable for the task (Zou et al., 2017). To computerize this work, two major activities are needed: (1) to formalize the building code and BIM into building rule models and building design representation models, respectively, and (2) to implement both models in computer programmes and execute rule objects over design objects in compliance checking automatically (Zou et al., 2017). Automation and optimization of the risks assessment process and decision flow is possible using specific plugins as Dynamo (Cortés, 2017). Some BIM platforms provide specific tools for a preliminary clash detection. The verification systems can be either a plugin for a given software, for example, Tekla BIMSight or Autodesk Navisworks or a stand-alone tool (e.g. Solibri).

Several authors have already conducted research on using rule-checkers for OSH purposes:

- Wang studied risk analysis automation with visualization of risk preventive measures as output (Wang et al., 2004). In 2011, Zhang introduced the first preliminary results of an automated safety rule checker for BIM using OSHA rules and Tekla Structures focusing on fall protection from leading slab edges, slab holes, and wall openings, automatically modelling the suggested preventive measures at the same time (Zhang et al., 2011). Benjaoran used, in a 4D model case study, a rule-based system to identify hazardous situations in work at heights as well as to establish preventive measures as part of 4D planning (Benjaoran and Bhokha, 2010). Taiebat has defined a framework for hazard identification in BIM, which supports the development of a BIM-based hazard recognition tool (Taiebat, 2011). Qi designed a prototype for safety checking to automatically identify fall hazards in a BIM model and provide design alternatives (Qi et al., 2011). Shih explored a BIM-enabled rule-checking system to help identify and mitigate OHS risks (Shih et al., 2012). Melzner compared rule implementation for risk assessment in the United States and Germany (Melzner et al., 2013). Sulanikivi researched risk identification during early planning phases and developed a rule-checker which was tested in a residential building project in Finland (Sulankivi et al., 2013). Kim proposed the automation of scaffold design and planning (Kim and Teizer, 2014). Zolfagharian proposed an automated safety planning plugin for scheduling software applications whose objective is to mitigate the occurrence of construction accidents (Irizarry et al., 2014). Qi focused on addressing fall risks in the design phase, and tried a safety rule translation algorithm to create machine-readable rules with Solibri (Qi et al., 2014). Park focused on emergency issues in case studies and proposed a checking process for analysing emergency evacuation egress routes and evacuation safety zones (Park and Kim, 2015). A rule-checking tool for managing fall and cave-in hazards in excavation pits was suggested by Wang (Wang et al., 2015). Zhang, in a practical case in Finland, focused on the identification and mitigation during planning of fall from heights risk while also considering preventive measures (Zhang et al., 2016). Wang created a rule checking tool for identifying fall and cave-in hazards in excavation activities (Wang et al., 2015). Zhang proposed an ontology-based semantic BIM modelling which enables inquiry of safety knowledge and analysis of construction site hazards (Zhang et al., 2014). Takim created "Automated Safety Rule Checking (ASRC)" to analyse falling hazards (Takim et al., 2016). Hongling et al. (2016) performed the integration of design safety codes and OSHA regulations with BIM to automatically identify safety issues (Hongling et al., 2016). Ding suggested an ontology-based methodology/framework in BIM environment for construction risk knowledge management and reuse in hope of indirectly improving the construction risk analysis process (Ding et al., 2016). Sadeghi studied BIM application through automated rule-checking for improving scaffolding systems and potential fall hazards (Sadeghi et al., 2016). Getuli proposed a validation workflow, through a parametric ruleset, for establishing minimum requirements for construction site layout submission (Getuli et al., 2017). Hossain focused on rule-based knowledge database using case studies (Hossain et al., 2018). Mihic studied the integration of BIM with a construction hazards database, to enable early hazard detection (Mihić et al., 2018). Wang used point cloud to check conformity of scaffolds, that is, the existence of handrails (Wang, 2019). Schwabe studied rules for verifying site planning (Schwabe et al., 2019). Hossain implemented an automated fall safety checking system in the planning phase (Hossain and Ahmed, 2019). Yuan focused on the design phase and created a rule-based inspection plugin for risk identification and pre-control measures (Yuan et al., 2019). Khan studied excavation planning through automated compliance checking (Khan et al., 2019). Teizer performed research on a safety-rule-checking platform – SafeBIM (Teizer and Melzner, 2018). Rodrigues created plugins "Job Hazard Analysis" and "Safeobject" to identify risks and apply preventive measures (Rodrigues et al., 2021). Shen studied a monitoring system for the construction of prefabricated buildings which leverages the potential of a plugin integration with BIM software for sharing, reusing, and accumulating knowledge regarding construction safety risk management in prefabricated building construction (Shen et al., 2020). Tariq developed a safety clauses visualization system to manage legal clauses of safety standards in an environment (Tariq et al., 2023).

So, BIM offers several advantages when used for the identification of hazards and risks including:

- 3D tools are more effective than 2D static drawings as they simulate actual working conditions which allows the visual assessment of these conditions

(Azhar and Behringer, 2013). See below, in Figure 24, 2D and 3D representations of same site;

- It increases the likelihood that hazards and risks will be detected in design phase (Mordue and Finch, 2019). Figure 3.1 exemplifies a task to be performed;
- It enables the clashes to be detected during the design phase (Mordue and Finch, 2019). Figure 3.2 illustrates a clash detection for geotechnical temporary support;

This is in accordance with European Directives about legal duties of designers in terms of OSH. It must be said that all these outputs rely on the assumption that designers are capable of making such an assessment or recognizing the possible

Figure 3.1 Risk identification through 3D (Kalloc Studios Asia Limited, 2024)

Figure 3.2 Clash detection (Tender, 2024)

need to undertake one. BIM, therefore, does not remove the need for professional expertise and construction knowledge.

- Design solutions and temporary works (Figure 3.3) can be configured and assessed without exposing workers to risk (Health and Safety Executive, 2018);
- Optimization of risk management (namely about the risks that could not be avoided in the design phase) through improved visualization (Azhar and Behringer, 2013) with a better perception, identification, and evaluation of risks, as well as scenarios of zones or time periods where there is a higher level of risks;
- Identification of potential constraints both in the work area and in the surroundings. Figure 3.4 illustrates constraints in construction area;
- Improvements in visualization and early simulation (Tender et al., 2018b) of actual working conditions (Azhar and Behringer, 2013), to identify, anticipate, and minimize risks before problems appear on the ground. Figure 3.5 illustrates the 3D visualization of a concrete structure;
- It enables better preventive measures and working conditions on the site and surrounding areas (Tender et al., 2022a);
- It provides suggestions to designers on how to alter their designs to make them safer to build (Mihić et al., 2019);
- 4D visualization to identify risks more effectively than conventional 2D (Hadikusumo and Rowlinson, 2004);
- Identification of preventive measures is carried out in a more automated way (Martínez-Aires et al., 2018) such as exemplified in Figure 3.6 where preventive measures concerning risk of soil collapse are shown;

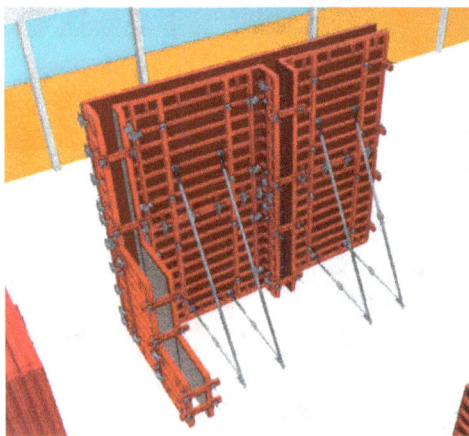

Figure 3.3 Temporary works design solution (PERI Group, 2024)

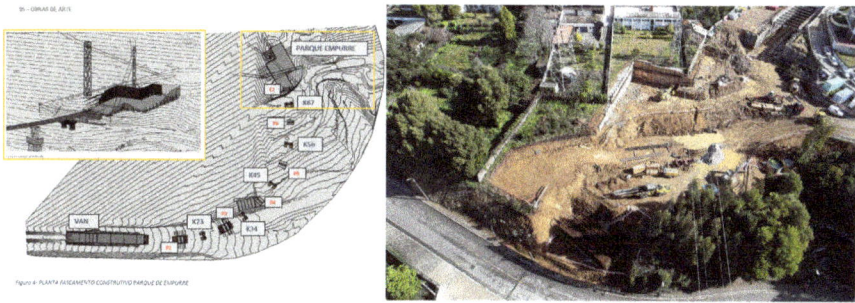

Figure 3.4 Constraints in construction area (Tender, 2024)

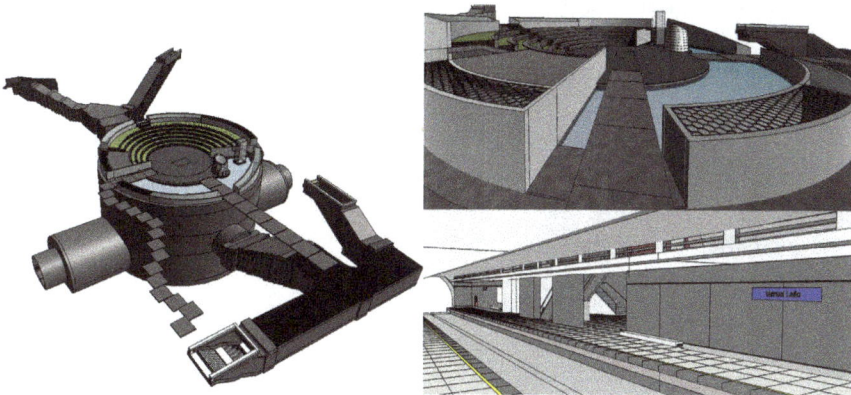

Figure 3.5 Concrete structure (Tender, 2024)

- Identification of risks can be made through symbols (Figure 3.7) that can be linked to documents or photos;
- Planning of delimitations to be settled (Figure 3.8);
- Planning of collective protections to be installed (guardrails, signalling networks, barriers; Figure 3.9);
- Clashes can be eliminated or mitigated quickly, reducing risk on site (Health and Safety Executive, 2018);
- Increased compatibility between specialties and elements, in the construction and the O&M phases, for example, the installation of collective protections such as guardrails or others;
- Establishment of safety perimeters for operations with mechanical cargo handling equipment (photo);
- It helps to manage constraints, namely aerial. Figure 3.10 is an example of aerial cable and its safety perimeter represented in 3D;

Figure 3.6 Temporary metallic structure (Tender, 2024)

Figure 3.7 Tunnelling boring machine 3D model with risk annotation (Ferrovial, 2024)

Figure 3.8 Delimitations (Kalloc Studios Asia Limited, 2024)

Figure 3.9 Guardrails in building (Tender, 2024)

- Use of BIM in the early stages of the project onwards (namely about techniques and equipment to be used – Figure 3.11) has been linked to a more effective connection to the production process (Azhar and Behringer, 2013);
- Increase in the performance of knowledge-based solutions by reducing flaws due to information exchange between building models and safety assessment tools (Fargnoli and Lombardi, 2020);
- Improved quality of safety coordination at the design stage (Mordue and Finch, 2019);
- BIM models enforce safety standards by monitoring compliance with regulations and identifying deviations from standards ensuring that the building is constructed in a safe manner;

Figure 3.10 Safety perimeter (Tender, 2024)

Figure 3.11 Excavation equipment (Kalloc Studios Asia Limited, 2024)

- Integration of safety planning and project planning allowing safety managers to be able to recognize when, where, and why safety measures on the safety plans must be used (Irizarry et al., 2014).

Rule-checkers prepared through BIM have the following advantages:

- While manual checking has the potential to be poor and prone to human error, BIM has the ability to allow this to be completed without errors;
- Verification and clash detection can be performed either by the designer, as a quality assurance aid, or by the recipient of the design;
- Identifies unsafe factors and safety hazards that in a conventional approach would probably not be noticed by OSH professionals;
- Eliminates human limitations in safety management (Hongling et al., 2016);
- Assists human decision-makers in safety planning to resolve the identified issues, by proposing realistic solutions (Hossain and Ahmed, 2019);
- Allows for rapid decision-making and evaluation (Park and Kim, 2015);
- Improves the quality of the design, in the design phase, by detecting errors and omissions (Park and Kim, 2015);
- Saves time and labour cost (Hongling et al., 2016).

Chapter 4

Training

4.1 Traditional approach

Training is considered as one of the most effective means of improving the safety performance of the construction industry (Li et al., 2012) and is an essential part of implementing accident prevention as it provides a strong foundation for general safety processes and requirements (Wetzel and Thabet, 2015). Conversely, a lack of training on OSH is one of the main causes for the large number of work accidents (Gonçalves et al., 2016). This shows how important training is to reduce the probability of having dangerous individual behaviours (Tender and Couto, 2017a). To comply with the legal requirement to provide training on OSH, companies adopt different systems both in the classroom and in the field (Bragança et al., 2020). Training is usually via lectures, videos, meetings, seminars, or demonstrations. In some countries, training is followed by mandatory assessment before being allowed to enter work sites (Tender and Couto, 2017a). However, "traditional" training methods often fail to achieve the desired objectives, due to their significant pedagogical limitations. They often lack properly defined evaluation methods, making it difficult to measure the effectiveness of safety training. This means that the intended behaviour change is difficult to measure, thus complicating implementation and mindset change in construction projects strongly driven by time and cost factors. Static in-person presentations, with no practical component, fail to engage workers, which means that operatives are not interested and cannot retain or apply critical safety concepts (Haslam et al., 2005). Training tends to be passive, which does not allow workers to test their understanding before engaging in the actual task during work (Dickinson et al., 2011). Training is often generic and does not reflect the complexities and specific spatial characteristics of the actual tasks. The transient and temporary nature of the workforce makes it difficult to implement new approaches. Language barriers and the time pressures of construction projects further exasperate the difficulties of transient workforces. Recorded videos can be used, adding a high degree of realism. However, these videos cannot be customized and do not always provide sufficient details for a given task of the particular project context.

DOI: 10.1201/9781003615217-4

Research has shown that the more time and effort an individual needs to spend obtaining information the less likely they are to retrieve the information and heed the stated warnings (Wetzel and Thabet, 2015). Training therefore seems to be an essential step towards improvement of construction safety (Akram et al., 2019).

4.2   BIM approach

Training has been the focus of some researchers:

Merivirta conducted a pilot study where LCD displays with 3D and 4D model views were placed in construction sites to present weekly updated safety information and optimize communication in construction site (Merivirta et al., 2011). Godfaurd proposed a framework for integrating BIM technology and planning software for health and safety site induction (Godfaurd and Abdulkadir, 2011). Ganah uses BIM for communication purposes (Ganah and Goudfard, 2015). Clevenger et al. (2015) developed and tested a bilingual construction safety training module using BIM-enabled visualization (Clevenger et al., 2015).

It is clear from such studies that BIM can be used as a training and information tool as it:

- uses visual communication (such as plans, sketches, photos, videos, and slide shows) which is one of the oldest and very high-impact ways to communicate. The first versions of writing were, after all, pictures (Merivirta et al., 2011);
- helps in organizing who attends training;
- provides visualization (Figure 4.1) which can help in the perception of how each task is carried out, as well as in the development of new ways of understanding the dangers (Mordue and Finch, 2019);
- provides faster access and dissemination of information (Tender et al., 2017b) with the ability to easily cross language barriers (Azhar and Behringer, 2013), thus improving training activities (Tender et al., 2018a) and communication between participants (Choe and Leite, 2017). BIM outputs can be more easily utilized for training purposes (Martínez-Aires et al., 2018). For example, a formwork to be used is represented in 3D (Figure 4.2);
- makes information available regardless of date or time of the day (Merivirta et al., 2011);
- increases the risk recognition capacity of workers, which makes the real-time communication between safety managers and workers more effective (Park et al., 2016), thus reducing the probability of accidents (Ganah and Goudfard, 2015);
- increases safety by making the site operatives more aware of what is going on there (Merivirta et al., 2011) (Figure 4.3);

Figure 4.1 3D workplace (Kalloc Studios Asia Limited, 2024)

Figure 4.2 Peripherical containance equipment (PERI Group, 2024)

- provides virtual walkthroughs, drawing attention to the use of the required personal protective equipment or other safety equipment for individual work processes;
- enables a faster and better means of conveying information (Tender et al., 2017b). Figure 4.4 exemplifies 3D of a reinforced concrete structure;
- improves the use and transmission of preventive information from the design phase to the construction phase which maximizes the optimization and interconnections between information (Mordue and Finch, 2019). Figure 4.5 is an

Figure 4.3 Activities on-going in site (Tender, 2024)

Figure 4.4 3D model (Tender, 2024)

Figure 4.5 Design of metallic temporary structure (Tender, 2024)

example of design information where the 3D visualization makes the design intent clearer;
- enables the visualization of global views and typically inaccessible site areas hidden from the naked eye, different conditions and building dimensions.

Chapter 5

Site planning

5.1 Traditional approach

The temporal and spatial approach to the management of the construction site is not always very systematic (Choe and Leite, 2017). Site material supply is important for the correct logistical management of the work (with the corresponding financial impact) due to its dynamic nature (Zhang et al., 2016) and the risks it creates for workers and for third parties (Tender et al., 2018b). Construction sites are made of several temporary facilities (office and welfare facilities, storage spaces, working areas, site infrastructures, lifting equipment, scaffolds, etc.) and corresponding accesses. The efficient management and layout planning of site facilities is an important factor contributing to successful construction management (Chau et al., 2004) and has been proven to reduce material handling costs while improving safety and productivity of a project (Kumar and Cheng, 2015). Traditionally these facilities are set up on unoccupied areas, within the boundaries of the construction site (Kumar and Cheng, 2015). Tendering documents typically do not consider these temporary facilities except for exceptionally complex temporary facilities such as cofferdams (Kim and Ahn, 2011). Most temporary facilities currently lack effective front-end planning and management (Kim and Teizer, 2014). In fact, as Sulankivi and colleagues report, detailed construction site layout planning is perhaps the best practical example of this paradigm where site safety issues are proactively studied and communicated in order to create working conditions where chances for accidents are minimized (Sulankivi et al., 2010). One of the most important factors that can adversely influence the OSH conditions on-site is the workspace layout (Getuli et al., 2017). The goal is to assure that the right materials, in the right quantities, at the right locations are provided at the right time to the construction crews on the project (Yu et al., 2016). Comprehensive site layout planning can ensure a smooth flow of materials, equipment, and labour, thereby improving the safety and efficiency of on-site operations (Kumar and Cheng, 2015). The temporary facilities and accesses have the potential to cause temporal and space conflicts which can imply loss of productivity or higher risks or even cause construction accidents, which

DOI: 10.1201/9781003615217-5

can be very serious (e.g. if a crane collapses). Site facilities management usually presents several challenges:

- site material supply (including special storage and delivery times) is almost always one of the major issues in construction site planning;
- the designers typically overlook safety considerations in temporary facilities design (Kim and Ahn, 2011);
- the site is "designed" only once before construction starts, without due consideration of the dynamic nature of supply issues (Yu et al., 2016);
- temporary facilities are generally not clearly delineated on the building drawings (Kim and Ahn, 2011);
- temporary facilities usually have to be moved during the construction;
- installation and dismantling of these facilities is one of the high-risk activities on worksites (Kim and Ahn, 2011);
- the positioning of temporary facilities entirely depends on the knowledge and experience of the site managers (Vimonsatit and Lim, 2014);
- although some construction plans include important temporary facilities late in the construction planning process, they are often installed at construction sites when needed but without sufficient planning effort (Kim and Teizer, 2014);
- manual site planning can be quite complicated, especially for projects with complex geometry and long schedules where changes to design or construction methodologies or sequences would have to be continuously updated into the site layout models, resulting in an inefficient workflow that is very time-consuming (Kumar and Cheng, 2015).

5.2   BIM approach

Several authors have studied the use of BIM for site planning:

Chau created a 4D model for planning daily activities more efficiently (Chau et al., 2004). Sulankivi worked on 3D and 4D models in order to manage site layout risk zones for crane operation (Sulankivi et al., 2010). Kiviniemi used 4D BIM models for managing and communicating construction safety plans (Kiviniemi et al., 2011). Kim proposed a prototype system for managing scaffolds in high-rise building projects (Kim and Ahn, 2011). New York City Department of Buildings (NYC DOB) published "BIM Site Safety Submission Guidelines and Standards" that enables contractors, through site plans electronic submission, to accomplish requirements of building permits also easing the compliance review process. At the same time, a standard library of site safety BIM objects has been made available, which contractors can use for site safety review. Vimonsatit showed how BIM tools can be used for construction site layout planning and simulation using 3D models (Vimonsatit and Lim, 2014). Kumar focused on estimation of temporary facilities and access ways during the construction phase, mainly in congested urban building projects (Kumar and Cheng, 2015). Trani presented a "Construction site information model" (CoSIM) with precise classification made on different types of elements (equipment, facilities, plants) according to their function on the construction site (Trani et al., 2015b). Trani developed a database for integration of construction site design in a design process, with several detail levels for BIM site objects (Trani et al., 2015a). Yi proposed including site safety planning in project schedule based on temporal and spatial inputs maximizing analysis of periods and zones in terms of safety (Yi et al., 2015). Cassano developed criteria for LOD scale for construction site temporary elements (Cassano and Trani, 2017). Biagini studied the impact of tower cranes in the neighbourhood in a rehabilitation site (Biagini et al., 2016). Yu developed a BIM-based dynamic approach for planning site material supply (Yu et al., 2016). Choe proposed a proactive site safety planning framework based on temporal and spatial input by integrating activity safety data with a project schedule and a 3D model (Choe and Leite, 2017). Getuli established criteria for site plan submission (Getuli et al., 2017). Ji studied tower cranes collision risks through 4D model (Ji and Leite, 2018).

BIM has the potential to improve site planning in the following ways:

- BIM facilitates the understanding of the impact on external environment namely pedestrian traffic (Figure 5.1);
- A BIM model can serve as the basis for a digital twin of the construction site and equipment, empowering stakeholders with a multi-dimensional view that streamlines planning, enhances safety, and optimizes logistics;
- BIM enables site layout planning and resource allocation, ensuring that the right resources are available at the right place and time, thus mitigating the risk of project delays and cost overruns (Tender et al., 2021). Figure 5.2 shows an example of 3D resource (material) allocation.

Figure 5.1 External impact of construction site (Kalloc Studios Asia Limited, 2024)

- BIM enables the generation of more illustrative site plans which are conducive to communication and collaborative decision-making (Sulankivi et al., 2010);
- the location of temporary facilities (Figure 5.3) can be assessed and managed more easily;
- the visual link between the schedule and construction site conditions (Figure 5.4) can facilitate decision-making during both the planning and the construction stages (Chau et al., 2004);
- more robust schedules and site layout and logistic plans can be generated to improve productivity (Kamardeen, 2010);
- A more detailed analysis of the construction site logistics is possible, which can be used to enhance the flows of materials and equipment (Figure 5.5) and optimize the space available at the construction site;
- risk zones, for example, safety perimeters (Figure 5.6), can be visualized through cylinders in 3D model, creating a better notion of what crane jib could hit (Sulankivi et al., 2010) and reducing hidden dangers, say, lifting injury accident, objects hitting, and collapses during the construction process (Yi et al., 2015);

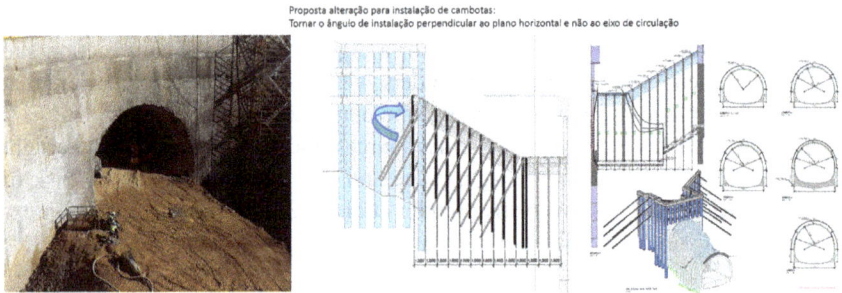

Figure 5.2 Tunnel portal layout plannning (Tender, 2024)

Figure 5.3 Location of temporary facilities (Tender, 2024)

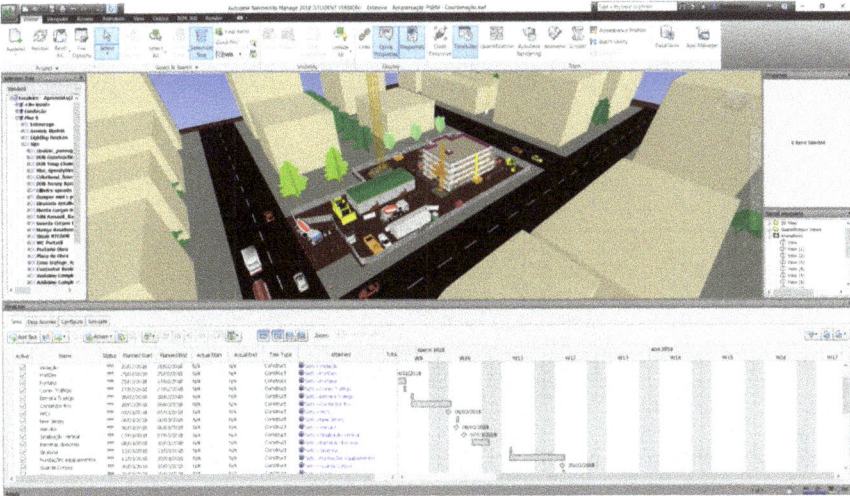

Figure 5.4 **Site arrangement through time (Tender, 2024)**

Figure 5.5 **Equipment necessary for tasks (Kalloc Studios Asia Limited, 2024)**

- BIM potentially facilities validation/approval by the PO or statutory bodies (Tender et al., 2018c);
- animations that can provide a quick general understanding of the site, and can be used, for example, as virtual sightseeing when introducing the project to site staff or to POs (Sulankivi et al., 2010) and stakeholders;
- real-time images and models (Figure 5.7) enhance communication between inspectors and supervisors (New York County Buildings, 2013);

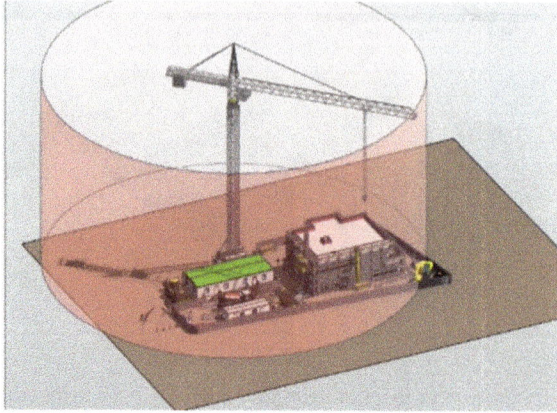

Figure 5.6 Crane safety perimeter

Figure 5.7 TBM model (Ferrovial Laing O' Rourke Joint Venture, 2016)

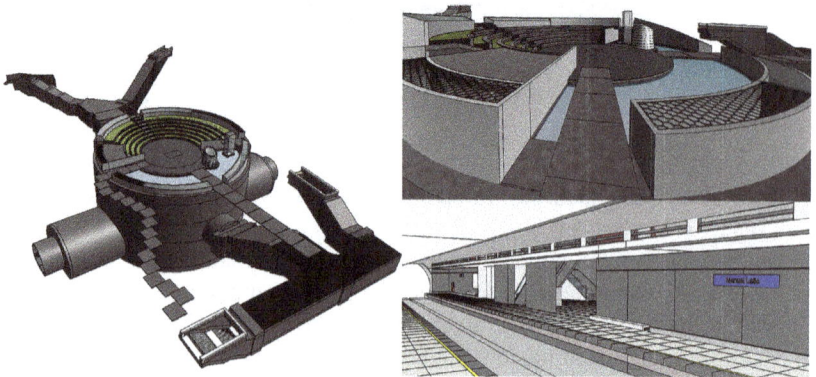

Figure 5.8 3D model of underground station (Tender, 2024)

- use of reality-capture technologies will allow sites to be explored in more detail, for example, 3D laser scanning or aerial drone photogrammetry;
- BIM offers an unprecedented ability, particularly in complex projects (Figure 5.8), to coordinate material, machinery, and human resources, leading to just-in-time delivery and reduced waste;
- Models can be used to run simulations and to check the suitability of welfare provisions and other site facilities (Health and Safety Executive, 2018).

Chapter 6

Task planning

6.1 Traditional approach

Shortcomings in traditional planning techniques which can lead to ineffective workspace management or schedule conflicts between activities are usually identified as major causes of high accidents rates. The planning of work, from the point of view of OSH, presents several difficulties including:

- OSH procedures generally are an excessive list of difficult-to-understand procedures, which are considered time-consuming by those who should be implementing them (Tender et al., 2017a). This limits the possibility of identifying and analysing the risks before the beginning of the construction phase (Azhar and Behringer, 2013);
- safety planning is usually undertaken in isolation from the construction planning, leading to poor connection between safety and work execution and planning (Azhar and Behringer, 2013);
- most usual planning techniques (Gantt chart, network diagram, and critical path method [CPM]) cannot account for the spatial feature of each activity and consider only construction schedules (Chau et al., 2004). Which can result in space-time clashes, or zones on a construction site that are temporarily congested, with multiple crews working on tasks that appear independent on the programme.
- traditional safety planning is carried out by manual means which are labour-intensive, tend to be error-prone, and are often highly inefficient (Getuli et al., 2017);
- the 2D approach to safety planning leads to a lack of accuracy in drawings (Getuli et al., 2017);
- reality is often subject to unexpected last-minute changes (which are usually unforeseen, unforeseeable, and accepted as part of a normal project work; Bargstädt, 2015), with new implications in terms of OSH;
- digital schedules are rarely updated frequently enough to accurately reflect all operations underway at any given point in time (Zhang et al., 2016), creating a lack of synchronization between high-level planning and daily operations;

DOI: 10.1201/9781003615217-6

- the absence of synchronization between high-level planning and daily planning complicates resource management, leading to inefficient allocation and utilization of labours, materials, and equipment on sites (Martins et al., 2022);
- unforeseen situations are usually accepted as part of a normal project work, and the managers/supervisors do not think or know they could be avoided by better anticipation.

It must be noted that safety planning plays an important role in construction project management for reducing unnecessary cost and delays related to undesired accidents (Chantawit et al., 2005).

6.2   BIM approach

Several authors have been debating the integration of OSH works planning through BIM:

Winch used BIM to explore and optimize critical zones in construction sites (Winch and North, 2006). Sulankivi focused on the implementation of safety measures in the form of guardrails as part of the model (Sulankivi et al., 2010). Eastman showed two practical examples of safety planning through BIM, related to safety perimeters (Eastman et al., 2008). Kim studied risk assessment in falsework objects focusing on scaffolding (Kim and Ahn, 2011). Chi developed specific BIM objects for temporary construction activities covering modularized scaffolding and formwork objects (Chi et al., 2012). Azhar studied BIM technology to perform safety planning and management in a case study focusing on crane management, excavation risk management, fall protection for leading edges, fall protection for roofers, and emergency response planning (Azhar et al., 2012). Zhang focused on automatically detecting falling hazards and setting preventive measures (Zhang et al., 2012). Sulankivi studied fall prevention planning in buildings (Sulankivi et al., 2013). Zhang focused on detecting workspace conflicts by analysing geometric conditions of different phases of the workspace to identify workspace congestion and safety hazards (Zhang et al., 2015). Berlo tested how the generation of limited and chosen drawings of specific tasks can have an impact on the knowledge of players in tasks (Berlo and Natrop, 2015). Bargstad focused on identifying challenges to site planning (Bargstädt, 2015). Ganah explored BIM for preventive purposes (Ganah and Goudfard, 2015). Kim focused on risks in temporary structures (Kim et al., 2016b). Feng studied risks in scaffold management using Dynamo (Feng and Lu, 2017). Wang investigated the implementation of scaffold safety (guard-rails and toe-boards) checking based on 3D point cloud data (Wang, 2019). Abed focused on fall and strike risks, based in 4D modelling (Abed et al., 2019).

BIM has the following advantages in terms of work planning:

- Workplaces can be separated and optimized (Figure 6.1);
- design solutions (Figure 6.2) can be conceived and assessed without exposing workers to risk (Health and Safety Executive, 2018);

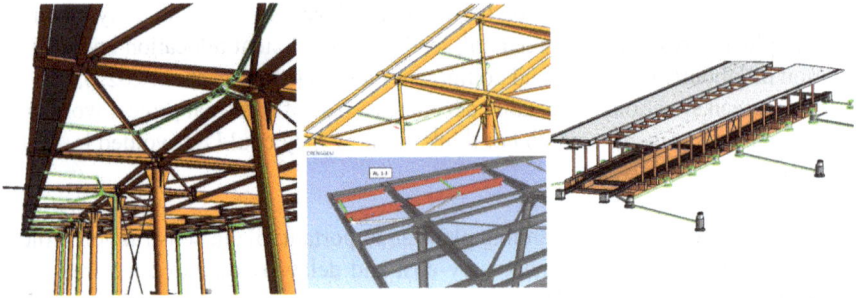

Figure 6.1 Workplace optimization (Kalloc Studios Asia Limited, 2024)

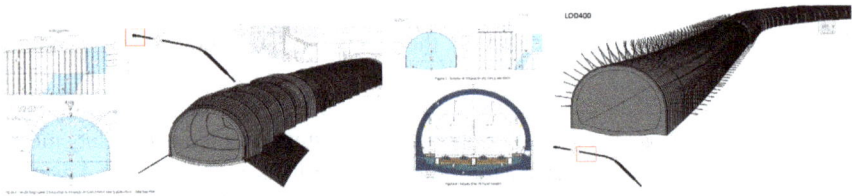

Figure 6.2 Design solutions for tunnel section (Tener, 2024)

- automated generation is possible of daily, weekly, or monthly updatable and changeable evacuation plans that reflect project progress, planned work schedules, and site conditions (Kim and Lee, 2019);
- BIM provides a detailed analysis of the construction site logistics (Figure 6.3) which can be used to enhance the flows of materials and equipment, optimize the space available at the construction site, and improve safety conditions (Zhang et al., 2016) in terms of the perception of risks by workers who are new to the construction site (Godfaurd and Abdulkadir, 2011);
- BIM can provide algorithms which automatically generate detailed schedules for temporary structures installation, namely scaffolding (Figure 6.4), which are not shown in the original contractor's schedule programme (Kim et al., 2016b);
- simulation allows a spatial and temporal analysis of tasks (Figure 6.5) and actual working conditions, optimizing planning, improving productivity, task sequence management, and material stack/equipment movement planning (Kim et al., 2016a);
- BIM streamlines the management of changes in materials, equipment, or workforce;
- benefits, risks, and costs of different phases (Figure 6.6) and solutions can be assessed and, if necessary, compared (Health and Safety Executive, 2018);
- clashes can be quickly eliminated or mitigated, reducing risks on-site (Health and Safety Executive, 2018);
- BIM reduces uncertainty of the temporary structures needed (New York County Buildings, 2013; Figure 6.7);

Figure 6.3 Equipment and materials in construction site (Kalloc Studios Asia Limited, 2024)

Figure 6.4 Site modelation (PERI Group, 2024)

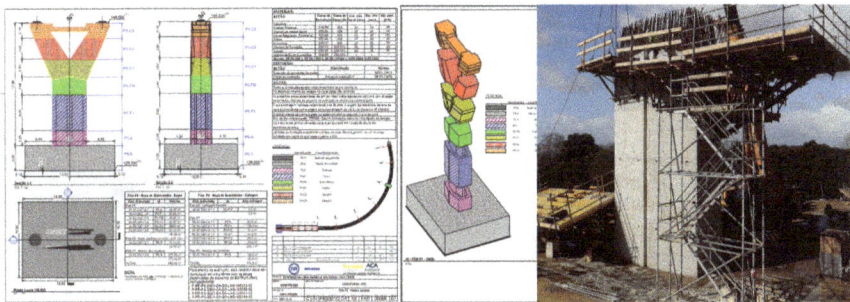

Figure 6.5 Phasing of pillar construction (Tender, 2024)

- BIM enables the visualization of potential hazards (Figure 6.8) before they manifest on-site;
- BIM helps to coordinate the location and timing of equipment use, preventing accidents related to equipment collisions and proximity to workers;

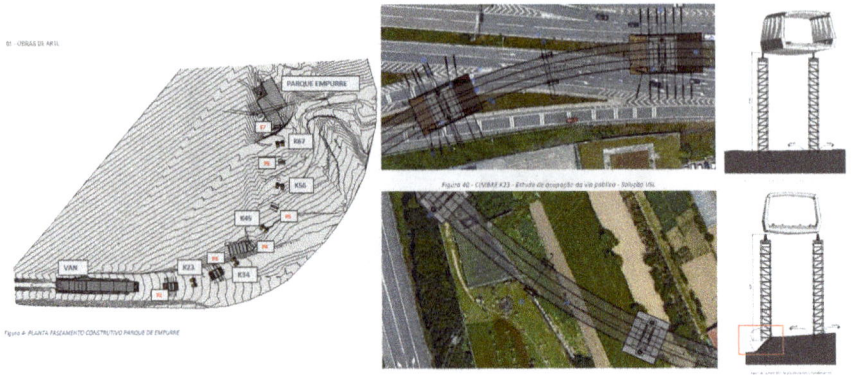

Figure 6.6 **3D models of metallic bridge (PERI Group, 2024)**

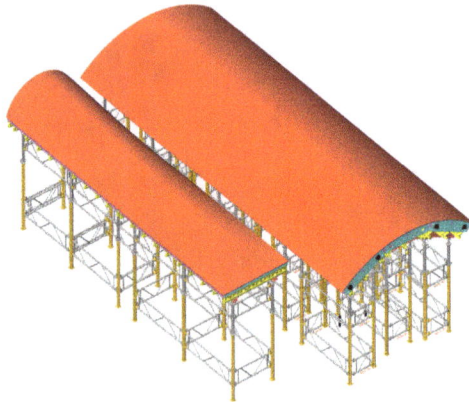

Figure 6.7 **Temporary structure (PERI Group, 2024)**

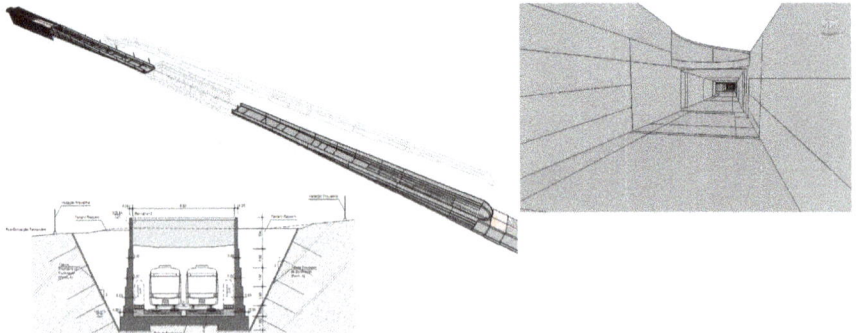

Figure 6.8 **3D models of metallic bridge (PERI Group, 2024)**

- BIM assists the planning of specific tasks (Figure 6.9), such as excavation, management of tower cranes, and work at height (Azhar et al., 2012);
- construction or installation sequences (Figure 6.10) can be modelled and assessed in terms of feasibility (Health and Safety Executive, 2018);

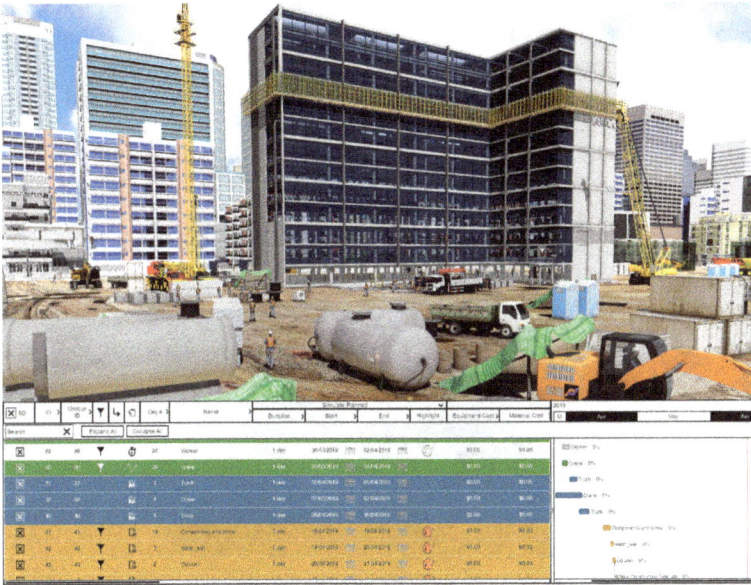

Figure 6.9 Task planning (Kalloc Studios Asia Limited, 2024)

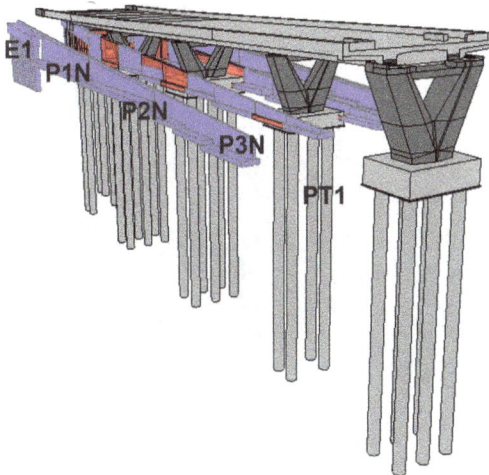

Figure 6.10 Construction sequence (Tender, 2024)

- workers can gain a better understanding of the project and its requirements, and can plan future activities more effectively (Health and Safety Executive, 2018);
- workers can provide input to improve the project plan and reduce risk (Health and Safety Executive, 2018);
- by simulating the construction sequence, BIM allows site managers to identify high-risk phases of construction and plan accordingly by instituting specific safety measures such as temporary guardrails (Figure 6.11) or netting systems during those times (Tender et al., 2021);
- BIM enables constructability of designs to be simulated and assessed at an early stage, leading to more effective task planning (Figure 6.12);

Figure 6.11 Guardrails and safety nets (Kalloc Studios Asia Limited, 2024)

Figure 6.12 Temporary formwork for concrete wall (Kalloc Studios Asia Limited, 2024)

- in cases of pre-fabricated elements (Figure 6.13), BIM can be used to ensure that the constructed parts can be manoeuvred into the correct position on-site (Health and Safety Executive, 2018);
- the feasibility of equipment, structures, and infrastructures can be tested before commitment (Health and Safety Executive, 2018);
- BIM facilitates collaborative working and speeds up the rate of decision-making (Health and Safety Executive, 2018);
- BIM improves integration of safety planning and project planning.

Several authors have been studying task planning through BIM. Zhang developed a 3D visualization model with schedule data (Zhang et al., 1998). Akinci created various types of space in 4D BIM for automated workspace conflict identification (Akinci et al., 2002). Chantawit used 4D to perform risk identification (Chantawit et al., 2005). Kiviniemi developed a detailed framework for fall protection modelling and 4D visualization, including the modelling of the temporary and permanent safety structures and equipment needed to carry out safe construction and O&M works (Chantawit et al., 2005). Hu focused on clash detection in the design phase, developing an algorithm for optimization of construction phase safety management (Hu et al., 2010b). Hu studied scaffold stability through 4D modelling (Hu et al., 2010a). Zhou purposed a prototype in which safety is integrated with schedules of site activities and uses automatic rule-based checking for identifying potential clashes before construction begins (Zhou et al., 2013). Kim studied automation of scaffolding objects in 4D BIM (Kim and Teizer, 2014). Choe formalized 4D construction safety planning process focusing on temporal and spatial safety

Figure 6.13 Earthworks equipment (Kalloc Studios Asia Limited, 2024)

information integration using a case study (Choe and Leite, 2017). Swallow researched the current opinion of professionals regarding the advantages and barriers for using 4D in safety management (Swallow and Zulu, 2019).

PAS 1192:6 stresses that "the use of 3D and 4D models in the design supports the principles related to an 'inherently safer design', to 'design safety' and to the legislative duties of the designers". Designers' views and attitudes towards risks are crucial to reducing them.

Planning with BIM, linking the 3D model with the schedule to create a 4D model, can have several positive impacts in OSH:

- The upfront simulation of the sequence of tasks leads to the early identification of various scenarios, with their opportunities, problems, constraints, and dangers, allowing to evaluate the advantages and disadvantages of each scenario (Mordue and Finch, 2019);
- Various OSH solutions and scenarios can be tested virtually, leading to better projection of financial costs which will assist in the optimization and control costs;
- A 4D model shows the construction sequence (Figure 6.14) from a safety stance, reducing the need for improvisation or last-minute solutions.

This is especially important in situations resulting from changes in the site organization that affect the work or the environment (Tender et al., 2018a) at critical times when there is an increase of the level of risks due to overlapping activities or in cases where there is a time or space incompatibility of tasks (Choe and Leite, 2017). Anticipating several scenarios has great potential for decreasing improvisation in site options (Martínez-Aires et al., 2018):

- helps the planning of the tasks, for example, constructability (which is typically not considered enough; Gilbertson et al., 2011) and the necessary resources for each one (Tender et al., 2017b). This can reduce the number of possible errors in the project (New York County Buildings, 2013) and facilitates the management of changes in materials, equipment, or labour;

Figure 6.14 Specific room for training (Kalloc Studios Asia Limited, 2024)

- locations can be associated with tasks, which is essential to be able to use specific planning techniques for OSH purposes;
- enables to recognize when, where, and why preventive measures on the safety plans must be used (Irizarry et al., 2014);
- identification of preventive measures (Figure 6.15) becomes more intuitive (Martínez-Aires et al., 2018);
- material requirements can be identified at any time and at any stage of the work as well as the estimated final cost related to each activity. This assists optimizing financial management (namely in relation to suppliers; Tender et al., 2018b) and improves the accuracy of estimates (Kamardeen, 2010);
- compatibility between specialities and general elements is increased, for example, the installation of collective protections (Tender et al., 2022a);
- assists the task of inspecting workplaces (New York County Buildings, 2013), allowing comparison between what is planned and what is done, within the scope of OSH, for example, setting up of collective protection;
- allows the swift translation of planning changes into safety-related changes (Tender et al., 2018b);
- enables a more structured, deeper, and less abstract approach to tasks, making the process closer to reality;
- better connection to the production process (Azhar and Behringer, 2013);
- reduced construction lifecycle cost, especially considering that the O&M phase lasts much longer than the design and construction phases (Matarneh et al., 2019);
- creates a safety planning practice that is undertaken earlier than traditionally in construction projects (Sulankivi et al., 2013);
- maximizes the capacity of sharing information (Tender et al., 2017b);

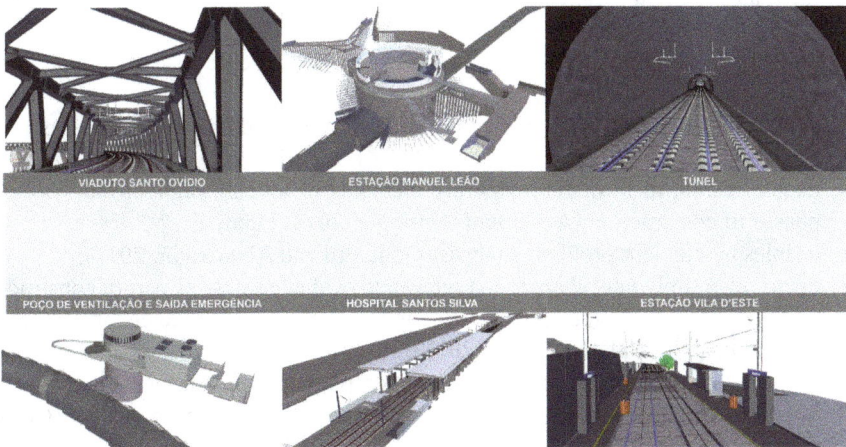

Figure 6.15 **Specific room for training (Tender, 2024)**

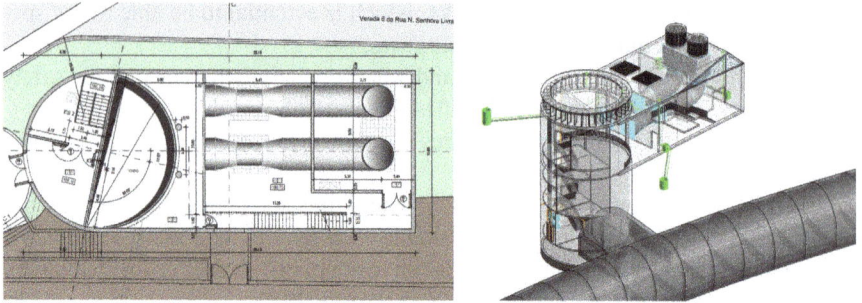

Figure 6.16 Ventilation shaft (Kalloc Studios Asia Limited, 2024)

Figure 6.17 Temporary equipment (Kalloc, 2024)

- facilitates communication with subcontractors and improves collaboration within the project team (Godfaurd and Abdulkadir, 2011);
- has a positive impact on safety awareness of site-specific hazards (Choe and Leite, 2017) (Figure 6.16);
- swift translation of planning changes into safety-related changes (Tender et al., 2018b);
- demonstration of how the temporary elements behave throughout the various phases of construction (Azhar and Behringer, 2013; Figure 6.17);
- enables an efficient problem analysis (Godfaurd and Abdulkadir, 2011);
- opens up entirely new chances to review and evaluate safety as part of construction operations to increase cooperation in safety planning and to enhance safety communication (Sulankivi et al., 2013);
- connects safety more closely to construction planning (Sulankivi et al., 2010);
- improves occupational safety by connecting safety issues more closely (Godfaurd and Abdulkadir, 2011).

Chapter 7

OSH monitoring

7.1 Traditional approach

The monitoring of correct OSH conditions in each task is currently one of the keys to efficient OSH management and is linked to the occurrence of accidents. It would be useful to have an excellent planning phase without a good monitoring and inspection system in place. To efficiently conduct safety monitoring, OSH experts conduct regular safety monitoring activities, such as daily inspections or regular audits, of construction sites. Based on these audits, they can identify and sanction unsafe acts and recommend further improvements. Monitoring depends on several factors, namely the size and complexity of the construction site, the resource assigned to monitoring (e.g. the number of safety technicians), and the type of documents that are used.

Additionally, site monitoring is traditionally based on text-based standalone checklists which are neither sufficiently timely nor adequate and accurate enough to successfully manage construction safety by identifying and recording breaches. It also relies heavily on manual observation, which is labour-intensive and error-prone. However, not all hazardous situations can be identified now when they first occur because construction sites can be large with many construction workers in different locations. Furthermore, H&S experts often have more than one construction site that they must supervise. The scale of most construction sites makes it impossible to monitor the whole site simultaneously without technology.

Safety monitoring is a critical step in the overall safety management process. It allows for making sure that planned measures are being implemented and provides insight into what measures work, what don't, and what can be done to further improve safety.

The risk management process is completed with the outputs of monitoring actions in the form of records, reports, or maps. These must be understandable by all parties and summarize the study carried out showing the results of the risk assessment and monitoring carried out (Longo, 2006), and are crucial for successfully conveying information.

DOI: 10.1201/9781003615217-7

7.2   BIM approach

Comparable to other previously mentioned OSH applications of BIM, BIM is not as often used for monitoring safety practices. This is understandable, given that in general, BIM use is more prevalent in the design phase, rather than in construction or O&M phases.

BIM by itself can be found used for the following purposes:

- visualization planned works as foreseen in safety procedures (Figure 7.1), for each work front (Mordue and Finch, 2019);
- allows comparison between what is planned and what is done, within the scope of OSH (e.g. setting up of collective protections);
- BIM-based monitoring enables the real-time compassion of actual construction progress against the construction plan (Mordue and Finch, 2019). This allows the work to be managed with greater focus on safety;
- BIM facilitates safety audit by seeing what activities are being performed and when and where, using spatial information from the BIM model and

Figure 7.1 Safety planning (PERI Group, 2024)

Figure 7.2 Dangerous working area (Kalloc Studios Asia Limited, 2024)

Figure 7.3 Scaffolding works (Kalloc, 2024)

temporal information from the BIM's fourth dimension, the integrated construction schedule;
- dangerous areas can be identified (Figure 7.2), which are affected by construction hazards that have wider area of effect, requiring urgent attention or more frequent monitoring;
- generating an automated report or checklist of hazardous activities that need to be audited, prior to going to the construction site;
- 4D BIM can be used to monitor and diminish the safety hazards that are associated with scaffolding work (Figure 7.3).

Chapter 8

Emergency planning and work accident investigation

8.1 Traditional approach

The first step in minimizing damage due to emergencies is to establish a contingency plan for each possible scenario (Tender et al., 2017b). This plan must consider the scenario of a possible catastrophe or accident by planning for first aid, rescue, and evacuation operations, which should be done in cooperation with external support entities.

One of the main focuses of this contingency plan is preparing emergency evacuation paths according to changing construction site configurations and construction progress. Several problems can occur: the number of workers on site can vary from day to day; site conditions can change every day; workspaces can clash with equipment paths; existing structures can block the pathway of workers. Since evacuees are prone to psychological tension in the process of emergency evacuation, reliable information sharing methods should be developed to help evacuees conveniently share relevant information (Deng et al., 2022). According to Marzouk, predicting evacuation time required for the labour to vacate is the major objective when preparing the evacuation plan for construction sites (Marzouk and Daour, 2018). Although it is surely crucial for workers' safety to prepare evacuation paths taking into account changing site configurations and construction progress, creating evacuation paths for all crews for each day can be an extremely labour-intensive task if it is done manually (Kim et al., 2016a). Nowadays, site managers use their past experience and intuitive understanding when planning moving paths of workers and equipment relying on heuristics or subjective judgement that can produce suboptimal or unsafe paths (Kim et al., 2016a).

The occurrence of AW or OD has a great impact on the work due to its effects on the temporal and financial planning of the work (Tender and Couto, 2016). Even though it is a negative indicator (Tender and Couto, 2017b), knowing the conditions in which an accident occurred or an OD arose has several advantages: (1) it provides an important basis for monitoring and improving preventive measures (European Commission – Eurostat, 2000); (2) it helps companies to comply with their legal requirements regarding risk assessment and to make better decisions that minimize their costs (Hale et al., 2007); (3) it allows learning from past mistakes

DOI: 10.1201/9781003615217-8

(Reis, 2007). It is important to use accidents as an indicator of OSH performance and to analyse them, but it is also important to pay attention to near-misses because they can help to predict future accidents (Marks and Shen, 2016). Not only accidents are important, so are near-misses that are "incidents where no property was damaged and no personal injury sustained, but where, given a slight shift in time or position, damage and/or injury easily could have occurred" (Marks and Shen, 2016). Near-misses are categorized as safety-leading indicators because they are measurements of processes and can predict future accidents.

8.2 BIM approach

Emergency planning using BIM has been studied by some authors:

Marzouk studied a framework for planning evacuation of workers, using BIM to estimate evacuation times (Marzouk and Daour, 2018). Kim implemented a framework to automatically analyse and generate the evacuation paths for multiple crews in 4D BIM (Kim and Lee, 2019).

BIM has the following potential advantages for the purpose of emergency planning:

- BIM can visualize the necessary safety scenarios and corresponding paths for egress during emergencies, ensuring clear routes are available and communicated to workers;
- in emergency situations a dispatched rescue team can quickly recognize the site conditions using the mapped 4D models (Figure 8.1) and determine the shortest and secure pathway to save labour from the site (Heaton et al., 2019);
- BIM can include the location of firefighting equipment (Tender et al., 2022a);
- in the event of an incident, the model could be made available to the fire and rescue services, who would be able to design their responses according to the location and availability of access routes and fire exits (Mordue and Finch, 2019).

In terms of accident investigation, some authors have already reported findings from research in this area:

Shen studied reports of near-misses using a specific BIM report database (Marks and Shen, 2016).

Faced with the risk of occurrence of an unfavourable event, such as a work accident or a near-miss, the potential of BIM can be used in several ways:

- to assist in the investigation of unfavourable events (Martínez-Aires et al., 2018) which allows for a better understanding of the tasks that were taking place, the tasks that were being carried out, the tasks which preceded the event (Azhar and Behringer, 2013), as well as to easily illustrate the flaws found (Azhar, 2011; Figure 8.1);
- BIM enables detection of incompatibilities between activities, allowing a reduction of near-misses and potential accidents (Mordue and Finch, 2019);

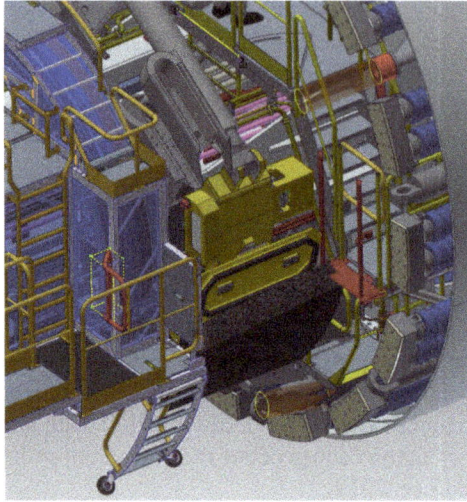

Figure 8.1 Illustration of unfavourable event scenario (Ferrovial, 2024)

- BIM helps OSH professionals to identify high-frequency and high-severity events within a construction site, for mitigation or hazard removal techniques (Marks and Shen, 2016);
- gives OSH professionals a simple way to identify locations where most near-misses are occurring and being reported (Marks and Shen, 2016).

Chapter 9

Operation and maintenance

9.1 Traditional approach

The Operation & Maintenance phase constitutes the most time-consuming phase, resulting in the highest cost during a project life cycle. On average, 85% of a project life cycle cost is concerned by the O&M phase (Azhar et al., 2012). Stakeholders involved in this phase have a much higher rate of injury and illness than the national average when compared to all other fields of employment. In the traditional approach, companies use several strategies to minimize risks, such as in construction phase. However, this information, coupled with the applicable safety information that is transferred from the design and construction phases (disconnect locations, power sources, O&M specific requirements, etc.) creates a myriad of fragmented data, presented in multiple resources, that must be reviewed or memorized in order to safety execute FM tasks (Wetzel and Thabet, 2018). Occupational health needs attention especially in both construction and O&M phases. Occupational Hazards in both phases are similar, however, both the risk probability and magnitude differ. In an attempt to mitigate safety hazards to O&M staff, a mechanism for transferring relevant safety information from the design and construction phase to the FM phase is of paramount importance (Haslam et al., 2005). For the O&M team, a complete understanding of the risk management process is of utmost importance given that, due to the operational and temporal requirements of O&M operations, there is a high risk of injuries, including electric shock, falls, crushing, cuts and bruises. Research has shown that the more time and effort an individual must spend obtaining information, the less likely they are to retrieve the information and obey the stated warnings; conversely, minimizing the amount of time and effort to the lowest possible level of information retrieval, has shown a much stronger likelihood of safety protocol implementation, this is especially important in a field where tasks are often time sensitive (Haslam et al., 2005).

9.2 BIM approach

It should be noted that O&M teams normally do not use BIM data models (including for the safety part), because these models, until recently, did not include the necessary O&M data, which makes the information exchange process tedious.

DOI: 10.1201/9781003615217-9

and overwhelming (Matarneh et al., 2019) and exacerbates the likelihood of work-related fatality, injury, or illness (Wetzel and Thabet, 2015). Furthermore, this information is rarely OSH-oriented. Exposure to risks such as noise, vibration, radiation, and abnormal thermal stresses are a significant occupational health concern that needs careful assessment and mitigation strategies to guarantee the OSH of building occupants. Recently, a significant push has been made to incorporate operations and maintenance information into BIM models for facilities maintenance and the linking to FM software. Through the use of BIM interoperability, virtual databases, and add-ons such as COBie, an improvement in O&M storage and retrieval has been achieved to some extent; however, these systems have still not evolved enough to handle all the interoperability. It is important to say that the final BIM model should not be merely a group of construction information elements, it is also necessary to guarantee that safety information is disclosed and presented (Kamardeen, 2010) as well as for a mechanism to emerge to transfer safety-relevant information from the design and construction phase to the FM phase. However, there is a lack of research focused on how BIM can be utilized to ensure worker safety by monitoring these parameters during the O&M phase. In the last decades, researchers targeted the use of BIM for OSH in the design, construction planning, and construction execution phases. Although the use of BIM in the O&M phase is not as well investigated (Hoeft and Trask, 2022), some authors have reported research results focusing on this:

Zhou have combined BIM and a domain ontology to facilitate the effective integration of dam safety monitoring information while reducing retrieval time effectively compared with traditional databases (Zhou et al., 2023). Wetzel developed a BIM-based framework to support safety during the facility management phase to categorize, consolidate, process, and present job specific, relevant safety information to FM personnel prior to the start of a maintenance task (Haslam et al., 2005). Liu propose a safety inspection method for water diversion projects using unmanned aerial vehicles and a dynamic BIM model which is created by aggregating timely-updated safety information with a BIM model in the Web environment (Liu et al., 2019).

BIM offers several advantages for FM phase:

- it enables a better understanding of the geometry of objects and conditionalisms (Figure 9.1)
- BIM provides a better facility monitoring through a powerful platform to store and process projects' information (Tang et al., 2019);
- BIM facilitates access to detailed information about the building's components, systems, and maintenance history, ensuring that safety-critical systems are properly maintained.
- Each material, component and facility can be linked to the appropriate element maintenance schedules, properties, manufacturers and other relevant information within the model (Figure 9.2), which means that once a part of the building requires maintenance, detailed information can be identified efficiently from the as-built model (Health and Safety Executive, 2018);

Figure 9.1 **3D representation of as-built (Tender, 2024)**

Figure 9.2 **3D model of railway equipment (Kalloc, 2024)**

- BIM can be used to track, update, and maintain facilities management information to support better planning, operations, and maintenance decision-making throughout a building's life cycle (Health and Safety Executive, 2018).
- BIM provides linked information about maintenance activities including maintenance schedules, conditions, equipment characteristics, and prospective risks;
- provides safety data, including hazardous equipment and material information;
- BIM allows detection of faults effectively by utilizing instrumental algorithms such as Fault Detection and Diagnostics that is able to identify damaged components;
- specific models can be developed and adapted for a particular end user, so that the information necessary for construction can be easily updated with details that will be of better use during maintenance (Mordue and Finch, 2019);
- BIM enables storage of the OSH information directly on the product, namely on its surface (Ventura et al., 2016), thus allowing improved recognition of

information during the life cycle, for example, by using barcodes or, in a more recent solution, QR Codes, linked to the central networks to manage maintenance needs and to improve the interaction between managers and maintenance teams, thus increasing the safety of the teams through 3D models.
- makes the O&M phase more effective in both sides: cost and time.
- BIM enhances the coordination and communication amid maintenance and safety teams.
- BIM offers significant advantages in detecting, reporting and responding to occupational hazards;
- The information contained within the as-built model can be used to plan and conduct future works after a construction is completed enabling hazards to be considered, reducing the risk to workers and members of the public (Health and Safety Executive, 2018).

BIM for OSH during O&M has emerged as a particularly important application. The Hackett Report published following the Grenfell Tower disaster in 2017 in the UK, calls for a "Golden Thread of Information" which tracks critical decisions throughout the building life cycle. This is intended in particular to record decisions made during the design phase, which would enable the building to be operated more safely. The Building Safety Act, which came into effect in the UK in 2022, further sets out the procedures and responsibilities for managing safety risks during the O&M phase. BIM clearly has a crucial role to play in complying with the Building Safety Act, and research on BIM for OSH during O&M is expected to grow.

Chapter 10

BIM and beyond

The built environment, as in most aspects of modern life, is being transformed by digital technologies. The concept of Construction 4.0 has been coined to coincide with Industry 4.0 (Porwal and Hewage, 2013). Industry 4.0 is distinct from earlier Industrial Revolutions in its emphasis on cyber-physical systems. Given the physicality of the built environment (in which physical 3D space is more important than in many other aspects of life), it is not surprising that the corresponding concept of Construction 4.0 has a particular resonance for the built environment.

Previous chapters of this book have dealt almost exclusively with BIM, only mentioning other digital technologies that can be used in construction OSH with BIM remaining as the central platform or visualization tool. This of course is one of the main features of BIM, to gather and store information in one place. Previous research that has conducted systematic (Chen et al., 2022) and bibliometric (Khudhair et al., 2021) reviews have found that BIM has been consistently used in combination with other technologies. BIM is often used as a starting point for other innovative technologies or used in cooperation with other technologies (Mihić et al., 2019), either as a data platform or for its graphical 3D representations. The ability to be integrated with various technologies comes with the ability to be used for various purposes. They of course can be useful on their own, but synergies can be realized when these technologies are combined with BIM.

Merging BIM with some of the other emerging or established technologies presented later can result in improvements to the construction process, but the uptake is slow due to the lack knowledge of these technologies, it being unclear whether they will have a positive or negative effect, and the lack of thorough investigation and comprehensive understanding on how they are complementing BIM and how they can help it reach its full potential. Kother technologies in combination with BIM can be utilized for various purposes in both design and construction stages. Safety monitoring can obviously be conducted only during the construction phase, hazard identification should be done during design and planning phases, but also during construction as new hazards emerge and safety training should be conducted before work starts. Due to the nature of the construction industry, simulation-based technologies are more applicable in the design stages, while those that are sensor-based are better suited to the construction phase.

DOI: 10.1201/9781003615217-10

The reason for BIM having this potential is that it can be characterized as a storage for various information and knowledge, which can be used throughout the project's life cycle and be valuable for overall project success (Steel et al., 2012), not only for H&S purposes. One of the problems is that BIM by itself can address some issues, but no one singular technology can solve all problems plaguing the construction industry. There are challenges to realization of such future, but they are out of the scope of this book.

As previously approached, Virtual Reality, Augmented Reality, the Internet of Things, Big Data, and Artificial Intelligence are going to be crucial. It will be important to understand how we can make the most of these new technologies and put them at the service of OSH as well as to study their weaknesses and threats detected to try to find suitable solutions.

These BIM-related technologies serve critical advantages to detect and prevent occupational hazards, including: (a) integrating project information (indoor spaces, equipment, components, documentations, safety rules, etc.), and hazard-related information (safety equipment location, emergency exit doors, hazardous areas, etc.) in a comprehensive 3D model. These increases recognizing and communication of safety measures; (b) connecting sensors to the model and enabling real-time data visualization; (c) detecting abnormal situations and its severity; (d) precisely locating and monitoring hazards; (e) tracking workers while identifying their respective locations; (f) providing workers, key players, and corresponding authorities with the necessary information; (g) aiding in decision-making and actions to mitigate hazards, such as shutting off equipment and sounding alerts. Their effectiveness is generally the same irrespective of BIM, but for others, such as location tracking, proximity detection and environmental sensing can greatly be improved by integrating location data from BIM. Let's take a look. Several authors have researched about using BIM with other digital technologies:

10.1   Virtual Reality (VR) and Augmented Reality (AR)

- to identify and visually display hazard information and improve safety knowledge (Park and Kim, 2013);
- it can be utilized for simulating high-risk situations and environment without putting workers at real danger;
- as an interactive educational tool in a virtual environment, allows to plan the works and identify safety risks during construction works and to propose measures for their elimination in virtual reality (Mesaros, 2023);
- development of virtual environments or sites to aid in safety training (Teizer et al., 2013);
- proactive behaviour-based safety approach, which focuses on delivering real-time alerts and conducting post real-time analyses for safety training by using automatic monitoring of workers' unsafe behaviours (Tender et al., 2022);
- simulating safety instructions on site for multilingual construction staff (Afzal and Shafiq, 2021);
- it can provide a realistic experience since this technology can respond quickly to user interactions and decisions; therefore it can be an effective tool for practising different work processes;

10.2   Gamification

Gamification can be used for safety training, in which trainees can learn and practice operating methods and construction sequences, which closely resemble the real working on-site environment (Guo et al., 2012). Workers would be able to walk through a virtual construction environment, with safety-related prompt messages appearing when they are close to unsafe operations (Li et al., 2012).

10.3   Internet of things

Sensors are probably the most common combination of BIM for safety monitoring. The term *sensor* to collect necessary data can encompass a lot of different sensors including vision-based sensors (cameras and kinetic sensors), Accelerometer-based (smart phones or wearing devices), and Radio Frequency sensor-based (Wi-Fi and radar system), photogrammetry and LIDAR. When BIM is linked to these sensors, it could provide informative data for (i) early detection, (ii) rapid response, (iii) effective mitigation of fire-related risks, (iv) ensuring the OSH of building users. Specifically, sensors can:

- enable the construction site to visually monitor the safety by collecting hazardous gas level and environmental condition (temperature and humidity) data and automatically alerting and starting ventilation on site if abnormalities are detected (Cheung et al., 2018);
- monitor a range of factors associated with fire hazards, including flame, heat, smoke, and other combustion products, air quality (presence of harmful particles), proximity detection and location tracking;
- Integrate an indoor positioning system inertial measurement unit, which can provide a worker's location and conditions, including their walking speed, and facing direction, so that the developed system could provide adequate safety notifications and real-time warnings based on safety regulations (Liu et al., 2020);
- be used for real-time tracking of personnel in order to generate real-time data to produce leading indicators for monitoring safety, security and verification (Costin et al., 2015);
- detect potential unsafe areas where workers are exposed to predefined risks and instantly communicate safety monitoring results are over the cloud for effective safety management (Park et al., 2016);
- allows a real-time access to O&M data (ii) maintenance checks, (iii) seamless creation and updating of digital assets, (iv) efficient space management, (v) identification of risks, (vi) reduction of false detection, (vii) alert notifications in BIM environment, and (viii) automation of equipment.

10.4   Artificial Intelligence (AI), Machine Learning

- could prevent the occurrence of hazards based on past experiences and self-learning process;
- monitor construction sites in real-time identifying safety violations, such as workers not wearing proper safety equipment or unsafe use of equipment and send alerts to supervisors.

- autonomous construction equipment, such as self-driving bulldozers and excavators, can be equipped with AI to ensure safer operations;
- alert supervisors if a worker is at risk of exhaustion or a health problem.
- create realistic scenarios for safety training where workers can practice responding to emergencies or hazardous situations in a safe, controlled environment before encountering them on actual construction sites.
- monitor safety compliance across construction sites identifying patterns of non-compliance and provide information for improving safety protocols.
- analyse large data sets to predict project delays, cost overruns and safety risks
- optimize emergency response plans by analysing site layouts, worker locations, and potential hazards.
- In the event of an emergency, can provide real-time evacuation guidance to ensure worker safety.
- to simultaneously display fire and evacuation simulation results offering precise information on the safest evacuation route, including the necessary evacuation distance and the specific emergency exit to take for a safe evacuation (Wehbe and Shahrour, 2021);
- fault detection to prevent fault hazard by monitoring equipment and hazard areas identifying the fault location and cause, identify the cause, activate protection systems, and alerts the first aid (Wehbe and Shahrour, 2021).

10.5 Wearables (hands free communication devices, such as smart helmets and vests equipped with sensors), big data

- worker's position and monitors worker's safety around a hazardous zone defined in the digital twin environment (Mun et al., 2023);
- collects real-time temperature and oxygen data remotely from wireless sensors placed at confined spaces on a construction site and notifies H&S managers with information needed to make decisions for evacuation (Arslan et al., 2014)
- Real-time dust monitoring system comprised of a network of low-cost mobile dust sensors and visualizations (Smaoui et al., 2018);
- workplace noise hazard prediction and visualization of noise spatial distribution (Wei et al., 2017);
- monitor conditions in real-time, such lightning levels (Teizer et al., 2017), structural weaknesses, gas leaks, or equipment malfunctions and send immediate alerts to prevent accidents.
- can track the workers' general health conditions such as heart rate, body temperature and body position to identify potential hazards before they occur (Awolusi et al., 2018);
- using beacons for data collection and trajectory analysis of worker movements through the construction site,
- data collection and trajectory analysis of worker movements through the construction site, which in turn is used for potential hazard identification (Arslan et al., 2019).
- detect potential unsafe areas, acquire real-time worker locations where workers are exposed to predefined risks and instantly communicate safety monitoring results are over the cloud for effective safety management (Park et al., 2016);

- localization of mobile construction resources (Fang et al., 2016);
- can warn that a worker needs to stop a certain activity for which a given threshold has been reached: for instance, stopping dynamic lifting (smart-digital-monitoring)
- Automated remote monitoring and assessing how the PPEs are worn through integrating pressure sensors and positioning technologies and through developing real time location system (RTLS) and virtual construction are developed for worker's location tracking (Dong et al., 2018);

10.6   UAV (Unmanned Aerial Vehicle) and UGV (Unmanned Ground Vehicle)

The sensor is not placed on a worker or a specific object, but rather is observing the point of interest from afar. They can be either static, mounted somewhere on a construction site, or mobile, carried either by a person or by an Unmanned Aerial Vehicle (UAV).

- can be operated remotely, reducing the need for operators to be physically present in dangerous areas;
- detect harmful environmental conditions such as poisonous dust in poorly ventilated areas, explosive gases, radiation leaks, chemicals and biological agents, and high humidity (smart-digital-monitoring)
- identifying construction equipment in construction site (Liang et al., 2018);
- conduct safety monitoring and provide inspection of the inaccessible, hard-to-reach, or unsafe locations on the site of a high-rise building construction site (Jhonattan et al., 2020);
- to streamline crane lifting operations planning and monitoring during the construction, by modelling the structure in BIM and capturing terrain data from UAV-captured images (Tian et al., 2021);

10.7   GPS, GIS and geofencing

- precise tracking of equipment and workers on construction sites;
- assist the optimization of the location of tower cranes (Irizarry and Karan, 2012);
- establish virtual boundaries and send alerts if equipment or workers enter restricted areas or hazardous zones;
- rapid identification of construction accidents and with prenotifications regarding safety hazards at a pertinent work zone by audio-based event detection (Lee et al., 2020);
- automated safety tracking mobile app with GPS to detect hazardous locations on construction sites, a cloud-based BIM system for visualization of worker tracking on a virtual construction site and a Web interface to visualize and monitor site safety (Hossain et al., 2023).

Chapter 11

Challenges and barriers to adoption

Anyone intending to adopt digital technologies in their OSH management system must conduct a thorough current and future needs assessment. The more reliable, timely, and exhaustive this assessment is, the greater its potential to respond to productive and competitive demands. The integration of digital technologies depends on the teams' ability to adopt new work processes, that is, professional need to embrace the new way of working and apply it in their activities. In fact, a new technological paradigm may be preceded by a phase of slow and gradual acceptance/implementation.

The introduction and implementation of BIM in the contracting processes inevitably requires fundamental changes and adaptations to the traditional processes that have been used for several years.

Significant progress is needed before BIM can be seen as an effective OSH information facilitator (Shih et al., 2012). The successful use of BIM requires a comprehensive understanding of its fundamentals, the relationship and potential of integration between new and traditional methods (Zou et al., 2017). If a company has already implemented BIM in other areas, it becomes simpler to create a path to integrate safety and its application in this area will be a natural consequence of the BIM adoption process (Mordue and Finch, 2019). After all, a new technological paradigm may be preceded by a slow and incremental acceptance and implementation phase. Despite its potential advantages BIM implementation often involves a variety of organizational barriers that can significantly influence intentions to use BIM purely based on technical or economic motivations (GhaffarianHoseini et al., 2017a) which can slow the implementation process. For BIM to promote rapid changes and overcome the cultural, institutional, and commercial barriers it is necessary that the organization's leadership and management teams recognize that the benefits of implementing BIM enough to justify its adoption. The motivations for implementing BIM within the companies' strategy are relatively complex and multidimensional (Cao et al., 2017). The implementation of BIM can be made for reasons that fall into three main categories: for image purposes, as a reaction to something, or for economic reasons. Image purposes and economic reasons are currently the strongest reasons for designers and contractors to implement BIM (Cao et al., 2017). Several reasons may justify the slower adoption of BIM, or even

DOI: 10.1201/9781003615217-11

its non-adoption. Despite its potential benefits, the implementation of BIM often faces different barriers depending on the context, which may have a significant influence on the intention of using it solely for technical or economic reasons, leading to a slower implementation. The prospect of BIM makes some organizations wary of the risk of a low or negative return on investment. It should be noted that the implementation of BIM is linked to the size of the company (Ghaffarian-Hoseini et al., 2017b). The obstacles and challenges are naturally greater in small and medium-sized enterprises and/or in small markets (where limitations may be exacerbated). Despite its various potential advantages for the OSH area, BIM presents a set of challenges and also barriers – which is perfectly natural when there is a change in processes and routines. Several authors studied the barriers and challenges for implementing BIM (e.g. Migilinskas, Ghaffarianhosein), some of them focusing on it for OSH purposes (e.g. Swallow (Swallow and Zulu, 2019)). The main barriers found were:

- **Not required by the** Project Owner – The Project Owner does not make its use compulsory. Several studies give this motive as the main barrier to BIM adoption. Consequently, it can be assumed that Project Owners requiring the use of BIM has significant implications for its widespread adoption. However, Project Owners generally do not require the use of BIM because they do not understand the technology and its advantages. Where Project Owners do not require the use of BIM, designers and contractors often struggle when trying to find arguments for BIM implementation (Tomek and Matejka, 2014). Olugboyega defended that what the Project Owners know about BIM is affecting what they are demanding in BIM (Olugboyega and Olugbenga, 2018). His research found that most of the Project Owners were not even aware of the whole concept of BIM and that the extra cost to support BIM use is one of the main reasons Project Owners refrain from demanding it.
- **Technicians' lack of training** – The lack of software and hardware training is another barrier identified in previous research. Academic and technical experts are scarce, mainly in BIM non-developed countries. Lack of training or skills to manage hardware and software can create delays in carrying out operational tasks. It often requires intervention, such as restarting training processes or addressing technical issues when they arise. Note that training usually requires more time of preparation and to become familiar with new technologies, for example, using VR headsets, in which younger workers have an advantage.
- it is important to increase the level of user-friendliness of the software environment as well as to make the hardware configuration as simple as possible;
- **Cultural resistance and reluctance to change** – The integration of BIM processes depends on the team's ability to adopt new work processes, that is, to absorb the new way of working and apply it in the delivery of their activities. They may prefer to keep doing the same thing for decades because they do not understand the benefits they can obtain by implementing BIM for OSH. Additionally, even when they understand its potential benefits, the fact that those

are not immediate may cause them to give up on its implementation. This kind of thinking may stem from mental states that are not open to change and from procedures that are based in oral communication instead of written. Behaviourally and culturally people may be reluctant to share information and as a result may show resistance to using BIM due to the way they perceive their work and because people tend to continue to use what works. Alomari stresses that sometimes even experienced safety managers are reluctant to use BIM technologies as they think that these cannot increase the project's safety level when compared to the traditional approaches (Alomari et al., 2017).

The minimal use of BIM as a data transference tool can be attributed to a number of issues. Issues such as model updates, a shortage of BIM skills by FM staff, a lack of collaboration between project and end user stakeholders, contract and legal framework, and interoperability, all contribute to the low utilization of BIM for FM (Becerik-Gerber et al., 2012). Arguably, the most complex issue faced by project teams, and where a significant amount of research has taken place, is *interoperability*. Within any project, various software platforms may be utilized to design and construct the facility, plan the work, store and exchange the information, and execute FM tasks.

- **Lacking knowledge of Return Of Investment (ROI)** – The first and most important step towards adopting any new technology is gauging its financial impacts (Akram et al., 2019) in terms of direct costs from buying hardware, software, upgrading IT systems, training for staff, and time for implementing procedures. There is a reduction in productivity when first implementing BIM in a company and this can often be unsustainable due to the high investment made and the resulting reduced productivity levels in the initial phases of adoption (Alomari et al., 2017). Usually, construction firms are hard-pressed to justify investment in new technologies as finance is scarce. ROI is one of the many ways to evaluate a proposed investment since it compares the anticipated gain from an investment to the cost of that investment (Azhar, 2011). However, the number of companies that are adopting BIM for OSH has been increasing and in most cases, this results in a considerable cost saving in the overall construction process. Based on several studies, the implementation of BIM can represent a viable investment and financial benefits can be achieved through BIM adoption (Eadie et al., 2013). One direct potential beneficial effect of BIM on safety is that while BIM might not directly address safety costs it helps in lowering the expenditure due to accidents as it is effective in improving construction safety (Akram et al., 2019). Based on this, it is recommended that several cost-benefit analyses be conducted to determine the best and most profitable way of implementing BIM in each particular company or project. The push-pull effect of the market will eventually reduce the costs and increase the accessibility for the rest of the industry to implement (Costin et al., 2018).
- **Interoperability problems** – Within any project different software applications can be used for communication management or for technical matters. Although IFC enables a common file, this has the potential to create a mismatch of file

names and extensions and proprietary systems that would not be able to communicate and work together (Clevenger et al., 2014). This issue is called *lack of interoperability* and is the main technological barrier to using BIM. It is essential to check translation (import/export information) between software applications with different data model structures.

- **Modelling detail**: the quality of the model effects the use of BIM for OSH, particularly in areas like training, which depend on the level of detail of the BIM model. The visualization of real construction situations provided by videos and photographs can only be achieved when the BIM model contains all important elements, like construction equipment, scaffolding and other temporary works. Additionally, the conditions during simulations are often hypothetical and simplified (Mordue, 2015); the translation of rules into machine readable codes requires significant manual work (Zhang et al., 2012); the different work processes (monitoring, training, coordination, etc.) often require heavy and expensive infrastructure (Park and Kim, 2015).
- **Using devices on site can create risks** – Using a device that enables the user to interact with vital construction information using electronic devices in a dangerous site environment may pose significant risks due to the possible distraction of the worker (Shah and Edwards, 2016).

Chapter 12

Conclusions

The aim of this book was to provide an updated view of the state-of-the-art relating to the use of BIM in OSH contexts in construction. After introducing the problem and outlining the application of BIM to OSH in Chapters 1 and 2, Chapter 3–10 presented specific OSH-related applications of BIM. The final chapters looked at technologies beyond BIM and explored challenges and barriers to adoption. The following specific conclusions may be drawn:

- Emerging digital technologies are revolutionizing the construction industry by offering unprecedented advantages and disruptive approaches.
- The opportunities for increased productivity, reduced timelines, improved safety, and sustainability are immense.
- Stakeholders who embrace these technological developments are poised to thrive in the evolving construction landscape. It is imperative for stakeholders to foster an environment that encourages innovation while simultaneously addressing these challenges to fully realize the benefits of digital transformation. As the AECO industry continues to evolve, it becomes clear that the digital transformation journey is as much about changing organizational cultures and mindsets as it is about adopting new technologies.
- As the industry strides towards more integrated and intelligent site management practices, BIM stands at the forefront, not just as a technology, but as a catalyst for a safer, more efficient, and more sustainable construction process.
- Digital technology options available are getting more advanced, capable, and affordable;
- Integrating OSH in digital data through BIM is a task that requires the engagement of all stakeholders, especially in scenarios characterized by the multidisciplinary nature of the work teams and a multi-organizational scope in construction projects.
- In recent years, there has been an increase in the understanding and application of BIM to construction with companies (namely large companies and multinationals) already making substantial efforts to implement BIM for OSH purposes in real-life scenarios.

DOI: 10.1201/9781003615217-12

- The implementation of BIM in companies requires a collaborative culture that people have the appropriate skills and are trained on the importance and usefulness of technology and that wider stakeholders have acquired some additional skills in relation to BIM
- Given the development of BIM, the mitigation and elimination of hazards and risks become more possible;
- The use of this methodology paves the way for better safety management systems with a greater speed and perception of prevention planning. It has great potential in terms of improving safety management and in the optimization of times and costs to increase production efficiency, providing a stronger link between production and safety.
- The results obtained show that this approach has the capability to revolutionize the integration of OSH information with digital data. It can improve the process of delivery, validation, and approval of OSH anchor documents. If successful, it will result in a paradigm shift that has great potential to reduce the occurrence of work accidents.
- The evidence to support the effectiveness of BIM for OSH management must be reinforced. Actors with successful BIM experiences need to disseminate their results, outcomes, and lessons learned to increase the awareness of prospective users/clients.
- The return on BIM investment in each individual case must be quantified as there is no "one size fits all" approach.
- SMEs are likely to face significant difficulties in the acquisition, implementation, and application of BIM for several reasons including cost, time, and availability of resources.
- Several factors (e.g. social, legal, financial, behavioural), which may change over the years, will affect the implementation of BIM in the OSH area. It is hard to make predictions about which applications will have a higher level of development or impact in the coming years. A continuous assessment must be performed.
- By enabling the desired reduction in accident rates and maintaining the total integrity of each worker, the needs of today's generations will be met without compromising the ability of future generations to meet their own aspirations which is one of the principles of sustainability.
- From BIM to drones, AI, and 5G connectivity, these technologies are transforming traditional construction practices, enhancing efficiency, safety, sustainability, and collaboration. BIM is just the beginning!

12.1   Future trends

One of the fundamental aspects of a book is its ability to open new ways for studies and research for the future. While researching this topic, secondary questions that cannot be explored in depth have been raised. It is the eternal question: after answering a question, more and more arise. As a result, several points that have been covered in less depth have great potential to become the focus of future studies.

BIM is not consolidated into current curriculum and in companies' training programs. Universities have a large responsibility in reinforcing this theme within each discipline.

The evolution of cities in the face of urban and demographic challenges has been extreme. What does the future hold? How will the construction sector be ten years from now is unknown and is a question that no one knows (nor dares to) the answer. BIM will evolve leveraged by its potential to provide a return on investment. However, we must be careful never to forget that we have to maintain a critical spirit and that technology is only a facilitator for the use and transmission of knowledge. This will not work without the correct education and mindset in terms of collaborative work (which we are so often reluctant to do).

It is also important for academic institutions to adapt to this new reality and needs to ensure that the preparation is done in classrooms by integrating this theme in all aspects of the AECO-related curriculum.

The exponential pace of technological advancement means that the book can become out of date before its writing has even been completed. Academia seems to be moving beyond BIM just as its adoption is beginning to gain momentum in industry. The technologies beyond BIM presented in Chapter 11 all represent fruitful avenues for future research. The digital transformation of construction means that most construction activities will now leave a huge trail of data, ripe for exploration using machine learning and particular branches of generative AI. It is an exciting time to be working in the digitally transformed construction sector. The workers of 18th-century Britain might not have realized that, three centuries later, they would become recognised as living in a pivotal period in the history of mankind. The same might be said of Construction 4.0 in the future.

Chapter 13

Useful information

13.1 Websites

Building Smart

www.buildingsmart.org

NBS

www.thenbs.com

B1M

www.theb1m.com

BIMPlus

www.bimplus.co.uk

UK BIM Alliance

www.ukbimalliance.org

BRE Group

www.bregroup.com

CDBB

www.cdbb.cam.ac.uk

UK BIM Framework

https://www.ukbimframework.org/

CIBSE

https://www.cibse.org/

BIM Corner

https://bimcorner.com/

DOI: 10.1201/9781003615217-13

13.2   Websites

BIM course

https://www.cursobim.com/

Post graduate in BIM Coordination

https://isep.ipp.pt/Course/Course/446

BIM A+ European Master in Building Information Modelling

https://bimaplus.org/

Master Internacional em BIM Management

https://www.e-zigurat.com

13.3   Software suppliers and development platforms

Kalloc

https://www.kalloc.com/home.html

Dynamo

https://dynamobim.org/

Python

https://www.python.org/

Solibri

https://www.solibri.com/

Bonsai

https://bonsaibim.org/

Acca

https://www.accasoftware.com/ptb/

That Open Company

https://thatopen.com/

OpenSource.Construction

https://opensource.construction/

This list is not comprehensive and, although the authors do not intend to recommend any particular commercial products, it is just for learning purposes.

References

Abed, H., Hatem, W. & Jasim, N. 2019. Adopting BIM technology in fall prevention plans. *Civil Engineering Journal*, 5.

Afzal, M. & Shafiq, M. 2021. Evaluating 4D-BIM and VR for effective safety communication and training: A case study of multilingual construction job-site crew. *Buildings*, 11.

Aguilera, A. 2017. Review of the state of knowledge of the BIM methodology applied to health. *In:* Arezes et al. (ed.) *Occupational Safety and Hygiene V*, Londres, Inglaterra: CRC Press/Balkema.

Ahmad, M., Demian, P. & Price, A. 2012. BIM implementation plans: A comparative analysis. *In:* 28th Annual ARCOM Conference, Association of Researchers in Construction Management, Edinburgh.

Akinci, B., Fischen, M., Levitt, R. & Carlson, R. 2002. Formalization and automation of time-space conflict analysis. *Journal of Computing in Civil Engineering*, 16, 124–134.

Akram, R., Thaheem, M., Nasir, A., Ali, T. & Khan, S. 2019. Exploring the role of building information modeling in construction safety through science mapping. *Safety Science*, 120, 456–470.

Ale, B., Baksteen, H., Bellamy, L., Bloemhof, A., Goossens, L., Hale, A., Mude, M., Oh, J., Papazoglou, I., Post, J. & Whiston, J. 2008. Quantifying occupational risk: The development of an occupational risk model. *Safety Science*, 46, 176–185.

Alizadehsalehi, S., Asnafi, M., Yitmen, I. & Celik, T. 2017. UAS-BIM based real-time hazard identification and safety monitoring of construction projects. *In:* 9th Nordic Conference on Construction Economics and Organization 13–14 June, 2017 at Chalmers University of Technology, Göteborg, Sweden, 22.

Alomari, K., Gambatese, J. & Anderson, J. 2017. Opportunities for using building information modeling to improve worker safety performance. *Safety*, 3.

Anbari, F., Carayannis, E. & Voetsch, R. 2008. Post-project reviews as a key project management competence. *Technovation*, 28, 633–643.

Aouad, G., Lee, A. & Wu, S. 2006. *Constructing the Future: ND Modelling*, London, United Kingdom: Taylor & Francis.

Arayici, Y., Egbu, C. & Coates, S. 2012. Building information modelling (BIM) implementation and remote construction projects: Issues, challenges, and critiques. *Journal of Information Technology in Construction*, 17, 75–92.

Arslan, K., Azhar, S. & Riaz, Z. 2014. Real-time environmental monitoring, visualization, and notification system for construction H&S management. *Journal of Information Technology in Construction*, 19, 72–91.

Arslan, M., Cruz, C. & Ginhac, D. 2019. Semantic trajectory insights for worker safety in dynamic environments. *Automation in Construction*, 106.

Awolusi, I., Marks, E. & Hallowell, M. 2018. Wearable technology for personalized construction safety monitoring and trending: Review of applicable devices. *Automation in Construction*, 85, 96–106.

Azhar, S. 2011. Building Information Modeling (BIM): Trends, benefits, risks and challenges for the AEC Industry. *Journal of Leadership and Management in Engineering & Science*, 11.

Azhar, S. & Behringer, A. 2013. A BIM-based approach for communicating and implementing a construction site safety plan. *In:* Associated Schools of Construction, ed. 49th ASC Annual International Conference Proceedings, Associated Schools of Construction, CA, EUA.

Azhar, S., Behringer, A., Sattineni, A. & Maqsood, T. 2012. BIM for facilitating construction safety planning and management at jobsites. *In:* CIB, ed. CIB W099 International Conference Modeling Building Health Safety, Singapura, 82–92.

Badri, A., Nadeau, S. & Gbodossou, A. 2013. A new practical approach to risk management for underground mining project in Quebec. *Journal of Loss Prevention in the Process Industries*, 26, 1145–1158.

Bargstädt, H. 2015. Challenges of BIM for construction site operations. *Procedia Engineering*, 117, 52–59.

Barnes, P. & Davies, N. 2014. *BIM in Principle and in Practice*, London: ICE Publishing.

Becerik-Gerber, B., Jazizadeh, F., Li, N. & Calis, G. 2012. Application areas and data requirements for BIM-enabled facilities management. *Journal of Construction Engineering Management*, 138, 431–442.

Behm, M. 2005. Linking construction fatalities to the design for construction safety concept. *Safety Science*, 43, 589–611.

Benjaoran, V. & Bhokha, S. 2010. An integrated safety management with construction management using 4D CAD model. *Safety Science*, 48, 395–403.

Berlo, L. & Natrop, M. 2015. BIM on the construction site: Providing hidden information on task specific drawings. *ITcon Vol. 20, Special Issue ECPPM 2014–10th European Conference on Product and Process Modelling*, 20, 97–106.

Biagini, C., Capone, P., Donato, V. & Facchini, N. 2016. Towards the BIM implementation for historical building restoration sites. *Automation in Construction*, 71, 74–86.

BIMSafe 2024. Private photo archive.

Bragança, D., Tender, M. & Couto, J. P. 2020. Building information modeling normative analysis applied to occupational risk prevention. *In: Occupational and Environmental Safety and Health II*, Springer.

The Business Research Company. 2021. *Global Construction Market Report 2021: COVID-19 Impact and Recovery to 2030*. https://www.thebusinessresearchcompany.com/press-release/global-construction-market-2021

Caires, B. 2013. *BIM as a tool to support the collaborative project between the structural engineer and the architect BIM execution plan, education and promotional initiatives.* MSc Dissertation, Integrated Master in Civil Engineering, University of Minho.

Cao, D., Li, H., Wang, G. & Huang, T. 2017. Identifying and contextualising the motivations for BIM implementation in construction projects: An empirical study in China. *International Journal of Project Management*, 35, 658–669.

Carvalho, F. & Melo, R. 2013. Stability and reproducibility of semi-quantitative risk assessment methods within the occupational health and safety scope. *Work*, 51, 591–600.

Cassano, M. & Trani, M. 2017. LOD standardization for construction site elements. *Procedia Engineering*, 196, 1057–1064.

(CDBB), C. F. D. B. B. 2018. *Year One Report towards a Digital Built Britain* [Online]. London. Available: https://www.cdbb.cam.ac.uk/Resources/ResoucePublications/CDB-BYearOneReport2018.pdf.

Ceyhan, C. 2012. *Occupational Health and Safety Hazard Identification, Risk Assessment, Determining Controls*. Tese de Mestrado em Engenharia Civil, Middle East Technical University.

Chantawit, D., Hadikusumo, B., Charoenngam, C. & Rowlinson, S. 2005. 4DCAD-safety: Visualizing project scheduling and safety planning. *Construction Innovation*, 5, 99–114.

Chau, K., Anson, M. & Zhang, J. 2004. Four-dimensional visualization of construction scheduling and site utilization. *Journal of Construction Engineering and Management*, 130, 598–606.

Chen, X., Chang-Richards, A., Pelosi, A., Jia, Y., Shen, X., Siddiqui, M. & Yang, N. 2022. Implementation of technologies in the construction industry: A systematic review. *Engineering, Construction and Architectural Management*, 29, 3181–3209.

Cheung, W., Lin, T. & Lin, Y. 2018. A real-time construction safety monitoring system for hazardous gas integrating wireless sensor network and building information modeling technologies. *Sensors*, 18, 436.

Chi, S., Hampson, K. & Biggs, H. 2012. Using BIM for smarter and safer scaffolding and formwork construction: A preliminary methodology. *In:* CIB, ed. CIB WO99 International Conference on Modelling and Building Health and Safety, Singapore.

Choe, S. & Leite, F. 2017. Construction safety planning: Site-specific temporal and spatial information integration. *Automation in Construction*, 84, 335–344.

Choi, B., Park, M., Lee, H., Cho, Y. & Lee, H. 2014. Framework for work-space planning using four-dimensional BIM in construction projects. *Journal of Construction Engineering and Management*, 140.

Clevenger, C., Glick, S. & del Puerto, C. 2015. Interactive BIM-enabled safety training piloted in construction education. *Advances in Engineering Education*, 4, n3.

Clevenger, C., Ozbek, M., Mahmoud, H. & Fanning, B. 2014. Impacts and benefits of implementing building information modeling on bridge infrastructure projects. *In:* Colorado State University (ed.), Project Brief | October 2014.

Cortés, A., Muriel, P. & Cortés, J. 2017. *Guía para la integración del subproceso coordinación de seguridad y salud en fase de diseño en el proceso de elaboración de un proyecto de edificación desarrollado con metodología BIM (Guide for the Integration of the Safety and Health Coordination Subprocess in the Design Phase in the Elaboration Process of a Building Project Developed with BIM Methodology)*, Universidad de Extremadura (University of Extremadura), Spain.

Cortés-Pérez, J., Cortés-Pérez, A. & Prieto-Muriel, P. 2020. BIM-integrated management of occupational hazards in building construction and maintenance. *Automation in Construction*, 113, 103–115.

Costin, A., Adibfar, A., Hu, H. & Chen, S. 2018. Building Information Modeling (BIM) for transportation infrastructure – literature review, applications, challenges, and recommendations. *Automation in Construction*, 94, 257–281.

Costin, A., Teizer, J. & Schoner, B. 2015. RFID and BIM-enabled worker location tracking to support real-time building protocol control and data visualization. *ITCon*, 20.

Couto, J. 2007. *Incumprimento de prazos de construção*. Tese de Doutoramento, Universidade do Minho.

Cruz, D. 2019. *Avaliação da aplicabilidade da realidade virtual como ferramenta de apoio à elaboração de planos de segurança*. Mestrado Integrado em Engenharia Civil, Instituto Superior Técnico, Lisboa.

Dainty, A., Leiringer, R., Fernie, S. & Harty, C. 2017. BIM and the small construction firm: A critical perspective. *Building Research & Information*, 45, 696–709.

Daniotti, B., Pavan, A., Spagnolo, S., Caffi, V., Pasini, D. & Mirarchi, C. 2020. *BIM-Based Collaborative Building Process Management*, Springer.

Delgado, J., Oyedele, L., Demian, P. & Beach, T. 2020. A research agenda for augmented and virtual reality in architecture, engineering and construction. *Advanced Engineering Informatics*, 45, August.

Deloitte 2020. *What Key Competencies are Needed in the Digital Age? – The Impact of Automation on Employees, Companies and Education*. Switzerland: Deloitte Switzerland.

Deng, H., Wei, X., Deng, Y. & Pan, H. 2022. Can information sharing among evacuees improve indoor emergency evacuation? An exploration study based on BIM and agent-based simulation. *Journal of Building Engineering*, 62.

Deng, L., Zhong, M., Liao, L., Lai, S. & Peng, L. 2019. Research on safety management application of dangerous sources in engineering construction based on BIM technology. *Advances in Civil Engineering*, 3, 1–10.

Dickinson, J., Woodard, P., Canas, R., Ahamed, S. & Lockston, D. 2011. Game-based trench safety education: Development and lessons learned. *ITcon – Special Issue Use of Gaming Technology in Architecture, Engineering and Construction*, 16, 118–132.

Digital Built Britain 2015. *Digital Built Britain – Level 3 Building Information Modelling – Strategic Plan*, Digital Built Britain.

Ding, L., Zhong, B., Wu, S. & Luo, H. 2016. Construction risk knowledge management in BIM using ontology and semantic web technology. *Safety Science*, 87, 202–213.

Directorate-General for Internal Market, I., Entrepreneurship and SMES (European Commission) 2019. *A Vision for the European Industry until 2030 – Final Report of the Industry 2030 High Level Industrial Roundtable. In:* Comission, E., ed. European Commission, Brussels, Belgium.

Dong, S., Li, H. & Yin, Q. 2018. Building information modeling in combination with real time location systems and sensors for safety performance enhancement. *Safety Science*, 102.

Eadie, R., Browne, M., Odeyinka, H., Mckeown, C. & Mcniff, S. 2013. BIM implementation throughout the UK construction project lifecycle: An analysis. *Automation in Construction*, 36, 145–151.

Eastman, C. 1975. The use of computers instead of drawings in building design. *Journal of American Institute of Architects*, 63, 46–50.

Eastman, C., Teicholz, P., Sacks, R. & Liston, K. 2008. *BIM Handbook-A Guide to Building Information Modeling for Owners, Managers, Designers, Engineers, and Contractors*, NJ, EUA: John Wiley & Sons.

European Commission 1989. *Directive 89/391/EEC – European Framework Directive*, European Commission.

European Commission 2008. *Causes and Circumstances of Accidents at Work in the EU*, Luxembourg: European Commission.

European Commission 2019. *European Construction Sector Observatory – Building Information Modelling in the EU Construction Sector*, European Commission.

European Commission – Eurostat 2000. *European Occupational Diseases Statistics – Phase 1 Methodology*, European Commission.

Eurostat 2023. *Construction Production (Volume) Index Overview*. Available: https://ec.europa.eu/eurostat/statistics-explained/index.php?title=Construction_production_(volume)_index_overview.

Fang, Y., Cho, Y. K., Zhang, S. & Perez, E. 2016. Case study of BIM and cloud–enabled real-time RFID indoor localization for construction management applications. *Journal of Construction Engineering Management*, 142, 05016003.

Fargnoli, M. & Lombardi, M. 2020. Building Information Modelling (BIM) to enhance occupational safety in construction activities: Research trends emerging from one decade of studies. *Buildings*, 10.

Farrell, L., Corbel, C. & Newman, T. 2020. Literacy and the workplace revolution: A social view of literate work practices in Industry 4.0. *Discourse Studies in the Cultural Politics of Education*, 42(6), 898–912.

Feng, C. & Lu, S. 2017. Using BIM to automate scaffolding planning for risk analysis at construction sites. *In:* ISARC, ed. 34th International Symposium on Automation and Robotics in Construction, ISARC, Taipe, Taiwan.

Fernández-Solís, J., Porwal, A. & Lavy, S. 2020. Impact of BIM and LEAN practices on construction project performance: An empirical study. *Journal of Construction Engineering and Management*, 146(1). https://doi.org/10.1061/(ASCE)CO.1943-7862.0001743

Ferrovial 2024. Ferrovial private photo archive.

Ferrovial Laing O' Rourke Joint Venture 2016. *NLE Fatigue Management*, Ferrovial Laing O' Rourke.

Fine, W. 1971. Mathematical evaluations for controlling hazards. *Journal of Safety Research*, 3(4), 157–166.

Gafari, M. & Aminzadeh, R. 2015. Identify and analyze the risks involved in tunnel projects. *Current World Environment*, 10, 1102–1108.

Ganah, A. & Goudfard, J. 2015. Integrating building information modeling and health and safety for onsite construction. *Safety and Health at Work*, 6, 39–45.

Getuli, V., Ventura, S., Capone, P. & Ciribini, A. 2017. BIM-based code checking for construction health and safety. *Procedia Engineering*, 196, 454–461.

GhaffarianHoseini, A., Tookey, J., Ghaffarianhoseini, A., Naismith, N., Azhar, S., Efimova, O. & Raahemifar, K. 2017a. Building information modelling (BIM) uptake: Clear benefits, understanding its implementation, risks and challenges. *Renewable and Sustainable Energy Reviews*, 75, 1046–1053.

GhaffarianHoseini, A., Zhang, T., Nwadigo, O., Ghaffarianhoseini, A., Naismith, N., Tookey, J. & Raahemifar, K. 2017b. Application of ND BIM Integrated Knowledge-Based Building Management System (BIM-IKBMS) for inspecting post-construction energy efficiency. *Renewable Sustainable Energy Reviews*, 72, 935–949.

Gilbertson, A., Kappia, J., Bosher, L. & Gibb, A. 2011. *Research Report – Preventing Catastrophic Events in Construction*. Health and Safety Executive, Londres, Inglaterra.

Godfaurd, J. & Abdulkadir, G. 2011. Integrating BIM and planning software for health and safety site induction. *In:* RICS, ed. The Royal Institution of Chartered Surveyors International Research Conference, University of Salford, Inglaterra, RICS.

Goh, C. & Abdul-Rahman, H. 2013. The identification and management of major risks in the Malaysian construction industry. *Journal of Construction in Developing Countries*, 18, 19–32.

Gonçalves, J., Tender, M. & Couto, J. 2016. Construction workers' training: Contributions to a more effective prevention culture. *In:* Arezes et al. (ed.) *Occupational Safety and Hygiene IV*, Londres, Inglaterra: CRC Press/Balkema.

Guo, H., Li, H., Chan, G. & Skitmore, M. 2012. Using game technologies to improve the safety of construction plant operations. *Accident Analysis and Prevention*, 48, 204–213.

Guo, H., Yu, Y. & Skitmore, M. 2017. Visualization technology-based construction safety management: A review. *Automation in Construction*, 73, 135–144.

Hadikusumo, B. & Rowlinson, S. 2004. Capturing safety knowledge using design-for-safety-process tool. *Journal of Construction Engineering Management*, 130, 281–289.

Hale, A., Ale, B., Goossens, L., Heijer, T., Bellamy, L., Mud, M., Roelen, A., Baksteen, H., Post, J., Papazoglou, I., Bloemhoff, A. & Oh, J. 2007. Modeling accidents for prioritizing prevention. *Reliability Engineering and System Safety*, 92, 1701–1715.

Haslam, R., Hide, S., Gibb, A., Gyi, D., Pavitt, T., Atkinson, S. & Duff, A. 2005. Contributing factors in construction accidents. *Applied Ergonomics*, 36, 401–415.

Hayne, G., Kumar, B. & Hare, B. 2014. The development of a framework for a design for safety BIM tool. *In:* ASCE, ed. International Conference on Computing in Civil and Building Engineering, Florida, USA, 49–56.

Hayne, G., Kumar, B. & Hare, W. 2016. Utilizing BIM technologies in the development of a mixed media approach to health and safety. *In:* International Workshop on Computing in Civil Engineering, Osaka, Japan.

Health and Safety Executive 2018. *Improving Health and Safety Outcomes in Construction – Making the Case for Building Information Modelling (BIM)*, Health and Safety Executive.

Heaton, J., Parlikad, A. K. & Schooling, J. 2019. Design and development of BIM models to support operations and maintenance. *Computers in Industry*, 111, 172–186.

Hermanus, M. 2007. Occupational health and safety in mining – status, new developments, and concerns. *The Journal of the Southern African Institute of Mining and Metallurgy*, 107, 531–538.

Hoeft, M. & Trask, C. 2022. Safety built right in: Exploring the occupational health and safety potential of BIM-based platforms throughout the building lifecycle. *Sustainability*, 14(10). https://doi.org/10.3390/su14106104

Hongling, G., Yantao, Y., Weisheng, Z. & Yan, L. 2016. BIM and safety rules based automated identification of unsafe design factors in construction. *Procedia Engineering*, 164, 467–472.

Hossain, A., Abbott, E., Chua, D., Nguyen, T. & Goh, Y. 2018. Design-for-safety knowledge library for BIM-integrated safety risk reviews. *Automation in Construction*, 94, 290–302.

Hossain, M. & Ahmed, S. 2019. Developing an automated safety checking system using BIM: A case study in the Bangladeshi construction industry. *International Journal of Construction Management*, 22(1), 1–19.

Hossain, M., Ahmed, S., Anam, A., Baxramovna, I., Meem, T., Sobuz, H. & Haq, I. 2023. BIM-based smart safety monitoring system using a mobile app: A case study in an ongoing construction site. *Construction Innovation*, 25(2).

Hu, Z., Zhang, J. & Zhang, X. 2010a. 4D construction safety information model-based safety analysis approach for scaffold system during construction. *Engineering Mechanics*, 12.

Hu, Z., Zhang, J. & Zhang, X. 2010b. Construction collision detection for site entities based on 4-D space-time model. *Journal of Tsinghua University Science Technology*, 50, 820–825.

Irizarry, J. & Karan, E. 2012. Optimizing location of tower cranes on construction sites through GIS and BIM integration. *Electronic Journal of Information Technology in Construction*, 17, 351–366.

Irizarry, J., Zolfagharian, S., Ressang, A., Nourbakhsh, M. & Gheisari, M. 2014. Automated safety planning approach for residential construction sites in Malaysia. *International Journal of Construction Management*, 14, 134–147.

Jhonattan, G., Gheisari, M. & Luis, F. 2020. UAV integration in current construction safety planning and monitoring processes: Case study of a high-rise building construction project in Chile. *Journal of Management in Engineering*, 36.

Ji, Y. & Leite, F. 2018. Automated tower crane planning: Leveraging 4-dimensional BIM and rule-based checking. *Automation in Construction*, 93, 78–90.

Kalloc Studios Asia Limited 2024. Photo private archive.

Kamardeen, I. 2010. 8D BIM modelling tool for accident prevention through design. *In:* Egbu, C., ed. 26th Annual ARCOM Conference, Association of Researchers in Construction Management, Leeds, Inglaterra, 281–289.

KarimiAzari, A., Mousavi, N., Mousavi, S. & Hosseini, S. 2011. Risk assessment model selection in construction industry. *Expert Systems with Applications*, 38, 9105–9111.

Khan, N., Ali, A., Skibniewski, M., Lee, D. & Park, C. 2019. Excavation safety modeling approach using BIM and VPL. *Advanced Civil Engineering*, 15.

Khanzode, V. V., Maiti, J. & Ray, P. K. 2011. A methodology for evaluation and monitoring of recurring hazards in underground coal mining. *Safety Science*, 49, 1172–1179.

Khanzode, V. V., Maiti, J. & Ray, P. K. 2012. Occupational injury and accident research: A comprehensive review. *Safety Science*, 50, 1355–1367.

Khudhair, A., Li, H., Ren, G. & Liu, S. 2021. Towards future BIM technology innovations: A bibliometric analysis of the literature. *Applied Sciences*, 11.

Kim, H. & Ahn, H. 2011. Temporary facility planning of a construction project using BIM (Building Information Modeling). *Computing in Civil Engineering*, 416(77).

Kim, H., Lee, H.-S., Park, M., Chung, B. & Hwang, S. 2016a. Automated hazardous area identification using laborers' actual and optimal routes. *Automation in Construction*, 65, 21–32.

Kim, K., Cho, Y. & Zhang, S. 2016b. Integrating work sequences and temporary structures into safety planning: Automated scaffolding-related safety hazard identification and prevention in BIM. *Automation in Construction*, 70, 128–142.

Kim, K. & Lee, Y. 2019. Automated generation of daily evacuation paths in 4D BIM. *Applied Sciences*, 9.

Kim, K. & Teizer, J. 2014. Automatic design and planning of scaffolding systems using building information modeling. *Advanced Engineering Informatics*, 28, 66–80.

Kiviniemi, M., Sulankivi, K., Kahkonen, K. & Merivirta, M. 2011. *VTT Tiedotteita – Research Notes: BIM-Based Safety Management and Communication for Building Construction*, VTT.

Koutamanis, A. 2020. Dimensionality in BIM: Why BIM cannot have more than four dimensions? *Automation in Construction*, 114.

Ku, K. & Mills, T. 2010. Research needs for building information modeling for construction safety. *In: International Proceedings of Associated Schools of Construction 45nd Annual Conference*, Boston, MA.

Kumar, S. & Cheng, J. 2015. A BIM-based automated site layout planning framework for congested construction sites. *Automation in Construction*, 59, 24–37.

Labagnara, D., Martinetti, A. & Patrucco, M. 2013. Tunneling operations, occupational S&H and environmental protection: A prevention through design approach. *American Journal of Applied Sciences*, 10, 1371–1377.

Latham, S. M. 1994. *Constructing the Team*, Latham Report – Final Report.

Lee, Y., Shariatfar, M., Rashidi, A. & Lee, H. 2020. Evidence-driven sound detection for prenotification and identification of construction safety hazards and accidents. *Automation in Construction*, 113.

Leitch, M. 2010. ISO 31000: 2009 – the new international standard on risk management. *Risk Analysis: An International Journal*, 30, 887–892.

Lena, E. & Eskesen, S. 2006. Guidelines for tunnelling risk assessment. *In:* Association, I. T., ed. World Tunnelling Congress, International Tunnelling Association, Seoul.

Li, H., Chan, G. & Skitmore, M. 2012. Visualizing safety assessment by integrating the use of game technology. *Automation in Construction*, 22, 498–505.

Li, M., Yu, H. & Liu, P. 2018. An automated safety risk recognition mechanism for underground construction at the pre-construction stage based on BIM. *Automation in Construction*, 91, 284–292.

Liang, C.-J., Kamat, V. & Messana, C. 2018. Real-time construction site layout and equipment monitoring. *In:* Wang, C. E. A., ed. Construction Research Congress, American Society of Civil Engineers (ASCE), New Orleans, USA.

Liu, D., Chen, J., Hu, D. & Zhang, Z. 2019. Dynamic BIM-augmented UAV safety inspection for water diversion project. *Computers in Industry*, 108.

Liu, D., Jin, Z. & Gambatese, J. 2020. Scenarios for integrating IPS–IMU system with BIM technology in construction safety control. *Practice Periodical on Structural Design and Construction*, 25.

Longo, S. 2006. *Análise e gestão do risco geotécnico em túneis*. Tese de Doutoramento em Engenharia de Minas, Instituto Superior Técnico.

Lu, Y., Wu, Z., Chang, R. & Li, Y. 2017. Building Information Modeling (BIM) for green buildings: A critical review and future directions. *Automation in Construction*, 83, 134–148.

Mahdevari, S., Shahriar, K. & Esfahanipour, A. 2014. Human health and safety risks management in underground coal mines using fuzzy TOPSIS. *Science of the Total Environment*, 488, 85–99.

Malekitabar, H., Ardeshir, A., Sebt, M. & Stouffs, R. 2016. Construction safety risk drivers: A BIM approach. *Safety Science*, 82, 445–455.

Marefat, A., Toosi, H. & Hasankhanlo, R. 2019. A BIM approach for construction safety: Applications, barriers and solutions. *Engineering Construction & Architectural Management*, 26.

Marhavilas, P., Koulouriotis, D. & Gemeni, V. 2011. Risk analysis and assessment methodologies in the work sites: On a review, classification and comparative study of the scientific literature of the period 2000–2009. *Journal of Loss Prevention in the Process Industries*, 24, 477–523.

Marks, E. & Shen, X. 2016. *Near Miss Information Visualization Application for BIM – Final Report*, The Center for Construction Research and Training, Silver Spring, USA.

Martínez-Aires, M., López-Alonso, M. & Martínez-Rojas, M. 2018. Building information modeling and safety management: A systematic review. *Safety Science*, 101, 11–18.

Martins, S., Evangelista, A., Hammad, A. & Tam, V. 2022. Evaluation of 4D BIM tools applicability in construction planning efficiency. *International Journal of Construction Management*, 22.

Marzouk, M. & Daour, I. 2018. Planning labor evacuation for construction sites using BIM and agent-based simulation. *Safety Science*, 109, 174–185.

Matarneh, S., Danso-Amoako, M., Al-Bizri, S., Gaterell, M. & Matarneh, R. 2019. Building information modeling for facilities management: A literature review and future research directions. *Journal of Building Engineering*, 24.

Melzner, J., Zhang, S., Teizer, J. & Bargstädt, H. 2013. A case study on automated safety compliance checking to assist fall protection design and planning in building information models. *Construction Management Economics*, 31, 661–674.

Merivirta, M., Mäkelä, T., Kiviniemi, M., Kähkönen, K., Sulankivi, K. & Koppinen, T. 2011. Exploitation of BIM based information displays for construction site safety communication. *In:* CIB W099 Conference, CIB, Washington, DC, USA, 24–26.

Mesaros, P. 2023. Increasing construction safety through virtual reality. *In:* Al, F. E., ed. CIBW099W123 Digital Transformation of Health and Safety in Construction, Porto, Portugal.

Mihić, M., Ceric, A. & Završki, I. 2018. Developing construction hazard database for automated hazard identification process. *Tehnicki Vjesnik*, 25, 1761–1769.

Mihić, M., Vukomanović, M. & Završki, I. 2019. Review of previous applications of innovative information technologies in construction health and safety. *Organization, Technology and Management in Construction*, 11, 1952–1967.

Mordue, S. 2015. BIM for health and safety, it's a no-brainer. *In:* Autodesk, ed. Autodesk University, Las Vegas.

Mordue, S. & Finch, R. 2019. *BIM for Construction Health and Safety*, Routledge.

Mun, S., Son, S., Lee, C. & Lee, S. 2023. Development of construction site safety monitoring techniques using worker positioning and BIM. *Korean Society of Hazard Mitigation*, 23.

NBS 2021. *Digital Construction Report 2021*. Available: https://www.thenbs.com/digital-construction-report-2021/_download/NBS_digital_construction_report.pdf.

New York County Buildings 2013. *Building Information Modeling Site Safety Submission Guidelines and Standards*, New York: New York City Department of Buildings.

Nnaji, C. & Karakhan, A. 2020. Technologies for safety and health management in construction: Current use, implementation benefits and limitations, and adoption barriers. *Journal of Building Engineering*, 29(2).

Observatory', E. C. S. 2019. *European Construction Sector Observatory – Building Information Modelling in the EU Construction Sector*. Available: https://single-market-economy.ec.europa.eu/sectors/construction/observatory_en.

Observatory', E. C. S. 2023. *Analytical Report – Trends in the Construction Sector*. Available: https://single-market-economy.ec.europa.eu/sectors/construction/observatory/analytical-reports_en.

Oesterreich, T. & Teuteberg, F. 2016. Understanding the implications of digitisation and automation in the context of Industry 4.0: A triangulation approach and elements of a research agenda for the construction industry. *Computers in Industry*, 83, 121–139.

Olugboyega, O. & Olugbenga, A. 2018. Correlation analysis of benefits of building information modelling and clients' requirements. *Journal of Scientific and Engineering Research*, 5, 53–68.

Park, C. & Kim, H. 2013. A framework for construction safety management and visualization system. *Automation in Construction*, 33, 95–103.

Park, J., Kim, K. & Cho, Y. 2016. Framework of automated construction-safety monitoring using cloud-enabled BIM and BLE mobile tracking sensors. *Journal of Construction Engineering Management*, 143, 05016019.

Park, S. & Kim, I. 2015. BIM-based quality control for safety issues in the design and construction phases. *International Journal of Architectural Research*, 9, 111–129.

Pasman, H. 2015. Costs of accidents, costs of safety, risk-based economic decision making: Risk management. *In:* Owen, C. (ed.) *Risk Analysis and Control for Industrial Processes-Gas, Oil and Chemicals*, Waltham, EUA: Elsevier.

Perera, S., Ingirige, B., Obonyo, E. & Ruikar, K. 2017. *Advances in Construction ICT and e-Business*, Routledge.

Pérez, A., Muriel, P. & Pérez, J. 2017. *Guía para la integración del subproceso coordinación de seguridad y salud en fase de diseño en el proceso de elaboración de un proyecto de edificación desarrollado con metodología BIM*, Spain.

PERI Group 2024. Private photo archive.

Pidgeon, A. & Dawood, N. 2021. BIM adoption issues in infrastructure construction projects: Analysis and solutions. *Journal of Information Technology in Construction*, 26, 263–285.

Porwal, A. & Hewage, K. 2013. Building Information Modeling (BIM) partnering framework for public construction projects. *Automation in Construction*, 83, 134–148.

Procore. 2024. Available: www.procore.com [Accessed September 2024].

Qi, J., Issa, R., Hinze, J. & Olbina, S. 2011. Integration of safety in design through the use of building information modeling. *Computing in Civil Engineering*. https://doi.org/10.1061/41182(416)86

Qi, J., Issa, R., Olbina, S. & Hinze, J. 2014. Use of building information modeling in design to prevent construction worker falls. *Journal of Computing in Civil Engineering*, 28.

Reis, C. 2007. *Melhoria da eficácia dos planos de segurança na redução dos acidentes na construção*. PhD Thesis, Faculty of Engineering of University of Porto.

Rodrigues, F., Antunes, F. & Matos, R. 2021. Safety plugins for risks prevention through design resourcing BIM. *Construction Innovation*, 21, 1471–1475.

Rodrigues, M., Rubio-Romero, J., Arezes, P. & Soriano-Serrano, M. 2016. Occupational risk assessment at Olive Oil Mills: Limitations and new perspectives. *DYNA*, 83, 21–26.

Sadeghi, H., Mohandes, S. R., Hamid, A., Preece, C., Hedayati, A. & Singh, B. 2016. Reviewing the usefulness of BIM adoption in improving safety environment of construction projects. *Jurnal Teknologi*, 78.

Sardroud, J., Mehdizadehtavasani, M., Khorramabadi, A. & Ranjbardar, A. 2018. Barriers analysis to effective implementation of BIM in the construction industry. *In:* ISARC 2018–35th International Symposium on Automation and Robotics in Construction and International AEC/FM, Berlin, Germany.

Sawhney, A., Riley, M., Irizarry, J. & Riley, M. 2020. *Construction 4.0 – an Innovation Platform for the Built Environment*, Routledge.

Schwabe, K., Teizer, J. & König, M. 2019. Applying rule-based model-checking to construction site layout planning tasks. *Automation in Construction*, 97, 205–219.

Shah, R. & Edwards, J. 2016. Investigation of health and safety impact from the 'Site BIM' tools in the live construction sites. *Journal of Construction Engineering and Project Management*, 6, 1–7.

Shannon, H., Robson, L. & Guastello, S. 1999. Methodological criteria for evaluating occupational safety intervention research. *Safety Science*, 31, 161–179.

Shen, Y., Xu, M., Lin, Y., Cui, C., Shi, X. & Liu, Y. 2020. Safety risk management of prefabricated building construction based on Ontology technology in the BIM environment. *Buildings*, 12.

Shih, S., Sher, W., Gibb, A. & Smolders, J. 2012. BIM and OHS – designer and design coordinator adoption in the UK and Australia. *In:* CIB, ed. CIB W099 International Conference on 'Modelling and Building Health and Safety', Singapore.

Smaoui, N., Kim, K., Gnawaai, O. & Lee, Y. 2018. Respirable dust monitoring in construction sites and visualization in building information modeling using real-time sensor data. *Sensors and Materials*, 30.

Smith, P. 2014. BIM implementation – global strategies. *Procedia Engineering*, 85, 482–492.

Software, A. 2018. *The 7 Dimensions of BIM-3D, 4D, 5D, 6D, 7D BIM Explained* [Online]. Available: https://biblus.accasoftware.com/en/bim-dimensions/ [Accessed April 2023].

Steel, J., Drogemuller, R. & Toth, B. 2012. Model interoperability in building information modelling. *Software & Systems Modeling*, 11, 99–109.

Sulankivi, K., Kähkönen, K., Mäkelä, T. & Kiviniemi, M. 2010. 4D-BIM for construction safety planning. *In:* Proceedings of W099-Special Track 18th CIB World Building Congress, Manchester, 117–128.

Sulankivi, K., Zhang, S., Teizer, J., Eastman, C., Kiviniemi, M., Romo, I. & Granholm, L. 2013. Utilization of BIM-based automated safety checking in construction planning. *In:* CIB, ed. CIB World Congress, 05–09 May, CIB, Brisbane, Australia.

Swallow, M. & Zulu, S. 2019. Benefits and barriers to the adoption of 4D modeling for site health and safety management. *Frontiers in Built Environment*, 4, 1–12.

Taiebat, M. 2011. *Tuning Up BIM for Safety Analysis Proposing Modeling Logics for Application of BIM in DfS*, PHD, Virginia Polytechnic Institute and State University.

Takim, R., Zulkifli, M. & Nawawi, A. 2016. Integration of Automated Safety Rule Checking (ASRC) system for safety planning BIM-based projects in Malaysia. *Procedia – Social and Behavioral Sciences*, 222, 103–110.

Tang, S., Shelden, D. R., Eastman, C. M., Pishdad-Bozorgi, P. & Gao, X. 2019. A review of Building Information Modeling (BIM) and the Internet of Things (IoT) devices integration: Present status and future trends. *Automation in Construction*, 101, 127–139.

Tariq, A., Ali, B., Ullah, F. & Alqahtani, F. 2023. Reducing falls from heights through BIM: A dedicated system for visualizing safety standards. *Buildings*, 13.

Taylan, O., Bafail, A., Abdulaal, R. & Kabli, M. 2014. Construction projects selection and risk assessment by fuzzy AHP and fuzzy TOPSIS methodologies. *Applied Soft Computing*, 17, 105–116.

Teizer, J., Cheng, T. & Fang, Y. 2013. Location tracking and data visualization technology to advance construction ironworkers' education and training in safety and productivity. *Automation in Construction*, 35.

Teizer, J. & Melzner, J. 2018. Building information modeling – technology foundations and industry practice. *In:* Al., B. E. (ed.) *Building Information Modeling – Technology Foundations and Industry Practice*, Switzerland: Springer.

Teizer, J., Wolf, M., Golovina, O., Perschewsk, M., Propach, M., Neges, H. & König, M. 2017. Internet of Things (IoT) for integrating environmental and localization data in Building Information Modeling (BIM). *In:* 34rd International Association for Automation and Robotics in Construction, Taipe, Taiwan.

Tender, M. 2024. Private photo archive.

Tender, M. & Couto, J. P. 2016. "Safety and health" as a criterion in the choice of tunneling method. *In:* Arezes et al. (ed.) *Occupational Safety and Hygiene IV*, London, United Kingdom: CRC Press/Balkema.

Tender, M. & Couto, J. P. 2017a. Study on prevention implementation in tunnels construction: Marão Tunnel's (Portugal) singularities. *Revista de la Construcción (Construction Magazine)*, 16, 262–273.

Tender, M. & Couto, J. P. 2017b. Typification of the most common accidents at work and occupational diseases in tunnelling in Portugal. *In:* Arezes et al. (ed.) *Occupational Safety and Hygiene V*, London, United Kingdom: CRC Press/Balkema.

Tender, M., Couto, J. P., Baptista, J. & Garcia, A. 2016. Prevenção no Túnel do Marão. *Revista Segurança*, 11–18.

Tender, M., Couto, J. P. & Fernandes, J. 2017b. Using BIM for risk management on a construction site. *In:* Arezes et al. (ed.) *Occupational Safety and Hygiene V*, Londres, Inglaterra: CRC Press/Balkema.

Tender, M., Couto, J. P. & Ferreira, T. 2015. Prevention in underground construction with sequential excavation method. *In:* Arezes et al. (ed.) *Occupation Safety and Hygiene III*, London, United Kingdom: Taylor & Francis.

Tender, M., Couto, J. P. & Fuller, P. 2022a. Improving occupational health and safety data integration using building information modelling – an initial literature review. *In:* Al, A. E. (ed.) *Occupational and Environmental Safety and Health III*, Switzerland: Springer Nature.

Tender, M., Couto, J. P. & Fuller, P. 2022b. Integrating occupational health and safety data digitally using building information modelling – uses of BIM for OHS management. *In:* Al, A. E. (ed.) *Occupational and Environmental Safety and Health III*, Switzerland: Springer Nature.

Tender, M., Couto, J. P., Lopes, C., Cunha, T. & Reis, R. 2018a. BIM (Building Information Modelling) as a prevention tool in the design and construction phases. *In:* Al, A. E. (ed.) *Occupational Safety and Hygiene VI*, Londres: Taylor & Francis.

Tender, M., Couto, J. P., Reis, R., Lopes, C. & Cunha, T. 2017a. O BIM (Building Information Modelling) como instrumento de prevenção em fases de projeto, de obra e da manutenção do edificado. *Ingenium*, 161, 106–107.

Tender, M., Couto, J. P., Reis, R., Monteiro, P., Rocha, T., Delgado, T., Pinto, J. & Vicente, G. 2018b. BIM as a 3D, 4D and 5D management tool in a construction site. *In:* Azenha et al., ed. 2nd Portuguese BIM Congress, Technical Superior University, Lisbon, Portugal.

Tender, M., Couto, J. P., Reis, R., Monteiro, P., Rocha, T., Delgado, T., Pinto, J. & Vicente, G. 2018c. O BIM como ferramenta de gestão 3D, 4D e 5D de um estaleiro de construção (BIM as a 3D, 4D and 5D management tool for a construction site). *In:* Azenha, ed. 2° Congresso Português de BIM (2 Portuguese BIM Congress), Instituto Superior Técnico (Superior Technical Institute), Lisbon, Portugal.

Tender, M., Fuller, P., Couto, J. P., Demian, P., Chow, V., Silva, F., Reis, R., Reis, F., Vaughan, A. & Long, M. 2022c. Real world lessons that can assist construction organisations in implementing BIM to improve the OSH processes. *In:* Al, A. E., ed. PTBIM – Portuguese BIM Conference, Braga.

Tender, M., Fuller, P., Couto, J. P., Gibb, A. & Yeomans, S. 2021. Emerging technologies for health, safety and well-being in construction industry. *In:* Al, B. E. (ed.) *Industry 4.0 for the Built Environment – Methodologies, Technologies and Skills*, 369–390. Switzerland: Springer Nature.

Tender, M., Fuller, P., Demian, P., Chow, V., Silva, F. & Couto, J. 2023. *BIM4OSH Observatory: Central Repository to Monitor the Status of BIM Implementation for OSH – Purposed Architecture CIBW099W123*. International Conference – Digital Transformation of Health and Safety in Construction Porto, Portugal.

Tender, M., Fuller, P., Vaughan, A., Long, M., Couto, J., Demian, P., Chow, V., Reis, R., Reis, F. & Silva, F. 2022d. Lessons from implementation of Key Technological Developments to improve occupational safety and health processes in a complex UK-based construction project. *In:* CIB, ed. CIB W078 World Building Congress 2022, CIB, Melbourne, Australia.

Tian, J., Luo, S., Wang, X., Hu, J. & Yin, J. 2021. Crane lifting optimization and construction monitoring in steel bridge construction project based on BIM and UAV. *Advances in Civil Engineering*, 1(8), 1–15.

Tomek, A. & Matejka, P. 2014. The impact of BIM on risk management as an argument for its implementation in a construction company. *Procedia Engineering*, 85, 501–509.

Trani, M., Bossi, B., Cassano, M. & Todaro, D. 2015a. BIM oriented equipment choice on construction site. *Proceedings of International Structural Engineering and Construction*.

Trani, M., Cassano, M., Todaro, D. & Bossi, B. 2015b. BIM level of detail for construction site design. *Procedia Engineering*, 123, 581–589.

Turner, C., Oyekan, J., Stergioulas, L. & Griffin, D. 2020. Utilizing Industry 4.0 on the construction site: Challenges and opportunities. *IEEE Transactions on Industrial Informatics*, 99.

UK Government BIM Working Group – CDE Sub Group 2018. *Asset Information Management – Common Data Environment Functional Requirements*, UK BIM Working Group.

Ventura, C., Aroca, R., Antonialli, A., Abrão, A., Rubio, J. C. & Câmara, M. 2016. Towards part lifetime traceability using machined quick response codes. *Procedia Technology*, 26, 89–96.

Vimonsatit, V. & Lim, M. 2014. Use of BIM tools for site layout planning. *In:* Press, I. (ed.) *Sustainable Solutions in Structural Engineering and Construction*, ASEA-SEC.

Wang, H., Zhang, J., Chau, K. & Anson, M. 2004. 4D dynamic management for construction planning and resource utilization. *Automation in Construction*, 13, 575–589.

Wang, J., Zhang, S. & Teizer, J. 2015. Geotechnical and safety protective equipment planning using range point cloud data and rule checking in building information modeling. *Automation in Construction*, 49, 250–261.

Wang, Q. 2019. Automatic checks from 3D point cloud data for safety regulation compliance for scaffold work platforms. *Automation in Construction*, 104, 38–51.

Wehbe, R. & Shahrour, I. 2021. A BIM-based smart system for fire evacuation. *Future Internet*, 13.

Wei, W., Wang, C. & Y. Lee 2017. BIM-Based construction noise hazard prediction and visualization for Occupational Safety and Health awareness improvement. *Computing in Civil Engineering*, 262–269.

Wetzel, E. & Thabet, W. 2015. The use of a BIM-based framework to support safe facility management processes. *Automation in Construction*, 60, 12–24.

Wetzel, E. & Thabet, W. 2018. A case study towards transferring relevant safety information for facilities maintenance using BIM. *Journal of Information Technology in Construction*, 23, 53–75.

Winch, G. & North, S. 2006. Critical space analysis. *Journal of Construction Engineering Management*, 132, 473–481.

Xiahou, X., Dib, H., Yuan, J. & Tang, Y. 2016. Design for Safety (DFS) and Building Information Modeling (BIM): A review. *In:* International Conference on Construction and Real Estate Management, Alberta, Canada.

Yan, H. & Demian, P. 2008. Benefits and barriers of building information modelling. *In:* Press, T. U., ed. 12th International Conference on Computing in Civil and Building Engineering, Tingshua University Press, Beijing, China.

Yi, S., Zhang, X. & Calvo, M. 2015. Construction safety management of building project based on BIM. *Journal of Mechanical Engineering Research and Developments*, 38.

Yu, Q., Li, K. & Luo, H. 2016. A BIM-based dynamic model for site material supply. *Procedia Engineering*, 164, 526–533.

Yuan, J., Li, X., Xiahou, X., Tymvios, N., Zhou, Z. & Li, Q. 2019. Accident prevention through design (PtD): Integration of building information modeling and PtD knowledge base. *Automation in Construction*, 102, 86–104.

Zhang, J. P., Anson, M. & Wang, Q. 1998. A new 4D management approach to construction planning and site space utilization. *Computing in Civil Engineering*, 15–22.

Zhang, L., Wu, X., Ding, L., Skibniewski, M. & Lu, Y. 2016. BIM-based risk identification system in tunnel construction. *Journal of Civil Engineering and Management*, 22, 529–539.

Zhang, S., Boukamp, F. & Teizer, J. 2014. Ontology-based semantic modeling of safety management knowledge. *Computing in Civil and Building Engineering*, 146.

Zhang, S., Lee, J., Venugopal, M., Teizer, J. & Eastman, C. 2011. Integrating BIM and safety: An automated rule-based checking system for safety planning and simulation. *In:* CIB, ed. CIB W099 Conference: Prevention – Means to the End of Construction Injuries, Illnesses and Fatalities, Washington, DC.

Zhang, S., Teizer, J., Lee, J., Eastman, C. & Venugopal, M. 2012. Building Information Modeling (BIM) and safety: Automatic safety checking of construction models and schedules. *Automation in Construction*, 29, 183–195.

Zhang, S., Teizer, J., Pradhananga, N. & Eastman, C. 2015. Workforce location tracking to model, visualize and analyze workspace requirements in building information models for construction safety planning. *Automation in Construction*, 60, 74–86.

Zhang, Y., Liu, H., Kang, S. & Al-Hussein, M. 2020. Virtual reality applications for the built environment: Research trends and opportunities. *Automation in Construction*, 118.

Zhou, Y., Boo, T., Shu, X. & Li, Y. 2023. BIM and ontology-based knowledge management for dam safety monitoring. *Automation in Construction*, 145.

Zhou, Z., Irizarry, J. & Qiming, L. 2013. Applying advanced technology to improve safety management in the construction industry: A literature review. *Construction Management and Economics*, 31, 606–622.

Zou, Y., Kiviniemi, A. & Jones, S. 2017. A review of risk management through BIM and BIM-related technologies. *Safety Science*, 97, 88–98.

Index

Note: Page locators in *italic* indicate a figure, and page locators in **bold** indicate a table on the corresponding page.

For Product Safety Concerns and Information please contact our EU
representative GPSR@taylorandfrancis.com
Taylor & Francis Verlag GmbH, Kaufingerstraße 24, 80331 München, Germany

www.ingramcontent.com/pod-product-compliance
Lightning Source LLC
Chambersburg PA
CBHW070736220326
41598CB00024BA/3446